The Life of Plants

The Life of Plants

E. J. H. Corner

The University of Chicago Press
Chicago and London

The University of Chicago Press, Chicago 60637
The University of Chicago Press, Ltd., London

© 1964 by E. J. H. Corner
All rights reserved. Published 1964
Phoenix edition 1981
Printed in the United States of America

88 87 86 85 84 83 82 81   1 2 3 4 5

Library of Congress Cataloging in Publication Data

Corner, E. J. H. (Edred John Henry)
  The life of plants.

  Bibliography: p. 297
  Includes index.
  l. Botany. I. Title.
QK45.2.C67   1981   581   81-11436
ISBN 0-226-11586-0 (pbk.)   AACR2

Contents

List of Illustrations

Acknowledgements

I am deeply grateful to the following friends and colleagues for the loan of illustrations:

Mr Paul H. Allen, United Fruit Company, Boston, Mass., for the durians on the jacket; Mr Douglas Botting, Worcester Park, Surrey, England, for Plate 31 b, c; Dr A. Fahn, Hebrew University of Jerusalem, for Figs. 45, 46; Dr Bruce Hayward, New Mexico Western College, for Plate 32; Dr Walter Hodge, Kensington, Maryland, for Plates 9–11 and 30; Mr Edwin A. Menninger, Stuart, Florida, for Plate 14; Professor G. F. Papenfuss, University of California, for Plate 4 b; Professor R. D. Preston, F.R.S., University of Leeds, for Plate 2; Mr B. Smythies, Sarawak, for Plate 41; Professor T. E. Weier, University of California, for Plate 1; Professor W. Gordon Whaley, University of Texas, for Plate 3. I am indebted, also, to the Royal Society of London for Plates 5, 18 and 34, and to the airway-company Cruzeiro do Sul, Brazil, for Plates 7, 38 and 40, which were selected for me by Professor Hilgard O'Reilly Sternberg, University of Rio de Janeiro.

Preface

BIOLOGY has become dissection. A living thing is too complicated to be understood. It must be stripped down like some strange machine, and its parts removed, correlated, enlarged, analysed, and tested to see what they are made of and how they work. The microscope has been followed by the ultramicroscope and all the ingenious instrumentation of modern physics and chemistry. The complexity in structure revealed at one magnification is increased at a higher; the reactions discovered by specialists in one technique amplify without finality the discoveries of others. The biologist, searching into the mystery of life, probes more deeply and studies, of necessity, smaller bits and happenings. So the substance and the ways of the living are broken down, and from the pieces tomes are gathered in encyclopaedic summary; what is known of a cell, or of part of a cell, may fill a volume. It has been a puzzle therefore how to write a small book about plants *in toto*.

The solution is happily botanical. A vigorous tree puts forth many branches; its plums multiply but become smaller. Prune and there will be fewer and larger plums. I have lopped off most physics, chemistry, physiology, genetics, and anatomy until I am left with plant matter absorbing sunlight, making sugar, and exploiting this success. The plant can be thought of as an obstruction, as the blind man perceived from his stumblings who saw men as trees walking. The plant interferes not only with the passage of light but the flow of water and the creep of other things about it; thus it obstructs the continuum of inanimate nature. It is, however, an unusual obstruction. It evolves and comes to pervade the environment to the utmost possibility; it reproduces excessively in forms that inherit its increasing powers; and in the shadow of its reality we were brought forth. I follow, in offering this sketch of plant life, the spacious outlines contemplated a generation ago by an untravelled scholar of

Oxford, A. H. Church, and published in a thin, grey paper-back, long out of print, called *Thalassiophyta and the subaerial transmigration*; for in all my ventures no other work has helped me to enjoy so much the heritage of plants.

Thalassiophyta is the story of sea plants and it opens with the sentence, 'The beginnings of botany are in the sea'. It has not recruited botanists to oceanography; it has not weaned their textbooks from pond life; but it stresses the basic fact. Seaweeds begin their lives as microscopic units adrift in sea water. Unless they fix on to a rock or something firm they will not grow. Plants have two states: the free-living, which is minute, and the attached, which fulfils plant form. Besides the seaweed spores, sea water holds multitudes of similar microscopic plant-cells that never settle on to rocks or which, in doing so, die. At once appears the sketch of plant evolution. A primeval ocean of pre-Cambrian age, three thousand million years ago, evolved or at this distance one may say created the plant-cell. A primeval shore evolved the plant form, or what is usually taken to be the plant. Migration on to land adapted this form to the robust and long-lived tree. When the strangeness of seaweeds is overcome and their intricacies are familiar, land plants are understood as seaweeds selected from the great multiplicity of thalassiophytes, dried off on the surface, waxed against evaporation, rooted, piped for water flow, built up with the transparent bricks made from the excess of sugars, yet still reproduced by seaways. Through all this passage of time the reproductive cells have carried a 'memory' of where they have been and how they should behave, and according to these memories the young grows in suitable surroundings into the adult. An example of what this book must forgo is the science of these memories (genetics), which deals with the piecemeal effort to understand how the inherited memory is constructed. What must be emphasized is the limitation of plants and of animals according to their memories. A flowering plant may develop a thread-like vegetative system, resembling that of a lowly fungus, and a duckweed may resemble a liverwort in simplicity, but neither the fungus nor the liverwort has been through the past experiences that lead to the flowering plant; both are more limited in what they can do.

Sea and land have changed since plant life ran this course. The changes have brought the modern plants and animals, with which the primitive could not compete. If, as it seems, the ancient and prime evolution of plants is over, the whole progression was yet so long and involved and success so gradual that there was time and opportunity for the less successful, stuck for one reason or another at almost any stage of advancement, to adapt themselves to the modernizing world. These partial relics of the past, swept up in the great movement of living progress, are the means by which we can interpret the familiar picture of the oak with fern, moss, fungus, and microscopic alga growing on its trunk. To understand how plants are made and how they work is not enough; we wonder how they came. Some say we shall never know, but to travel botanically and to see the immense variety of plants that survive, and to realize how little yet is appreciated of them, fills me with optimism; only there must be a care lest civilization, which is the undoing of their wilderness, eliminates them.

The most baffling problem is how plants came on to land, and it is aggravated by the blurring of many modern situations. Advanced plants now mingle with the simple and have themselves been simplified. Lowly plants take short cuts into high society. In the inhospitable regions of the earth all kinds may find refuge. So plants of the simplicity of microscopic forms of the sea live with spongy mosses on mountains. Seagrasses mingle with seaweeds, and seaweeds float in sea water. Orchid, fungus, and seaweed grow on the roots of flowering mangrove trees. Coltsfoot and horsetail border the ploughed field. Baboons eat cycad seeds; fish eat tropical fruits. Trees need fungus. Through all this confusion shines Church's thesis.

Plant life is integrated with its surroundings. Its substance is fashioned by physical and chemical forces in the environment. Its form becomes the manner of obstructing or intercepting the goings-on. Every intercept makes new events. Then, as form enlarges, its influence extends and events multiply. Open sea is a simple situation for simple plants, whose floating part is primitive and microscopic. The shore gives purchase and the plant becomes twofold with holdfast and shoot; it starts to obstruct because it is attached, and it covers the substratum with vegetation, which obstructs the light and the flow of water; it builds the animate into the inanimate environment. The land introduces air as the third factor of the environment after water and rock. The more complicated conditions that ensue bring forth the more complicated forest, where at length seed, fruit, and flower predominate in structure and function far beyond the needs of the seaweed. Rigidity, longevity, and loftiness are the requirements of the land plant. The forest becomes the greatest living obstruction and in its emancipation brings forth the greatest delights of the vegetable world. Thus the three main scenes of the earth, which are the sea, the coast, and the land, define with their increasing complexity three main arenas of plant evolution in clarity far exceeding the pervasive course of the free animal.

Botany began with medicinal plants. Paper made the herbarium and the books. The microscope and the laboratory brought understanding. Now world exploration must rearrange the bits of knowledge into the story of plant life.

Plant anatomy and natural history, the critics declare! Where are the growing points of knowledge? Working back, I reply, as impetuous streams into the ancient mountain of life, while the rivers of thought grind the new facts into grains of knowledge. They are shifting over the delta of biology; its channels are becoming lost and it needs recharting. If the botanist can now see better over this delta, which is his outlet to the intellectual community, it is because, in the old simile, by sitting on the giant's shoulder the dwarf may see further than the giant himself. The books that deal with general botany have grown so tediously compendious, so canalized in the circuitous fertility, so thoroughly dull and dully thorough, that I have no hesitation in offering this survey, understandable, I hope, and therefore open to experiment, proveable, and progressive.

Chapter 1

The Ocean

A PLANT is a living thing that absorbs in microscopic amounts over its surface what it needs for growth. It spreads therefore an exterior whereas the animal develops, through its mouth, an interior. The definition is vague but it gives the reason why the limb of a tree ends in leaves, not fingers, and it helps to explain how in the long run of evolution a monkey came to sit on the tree and a cow to ruminate in its shade. Nevertheless, what the plant absorbs, how it grows, what it becomes, and how it reproduces are properties so various that seaweeds, fungi, and bacteria, as well as flowering plants, ferns, and mosses, must be enrolled in the plant kingdom. The problem is where to begin thinking about this host of vegetable life which thrives without intention, builds without circulating blood, feels and responds without sense-organs and muscles, summons animals without contriving, and serves in its over-production their food supply.

It might be thought that the life of plants could be gathered sufficiently from the familiar vegetation of the land and, in particular, from the objects of field and garden. But these plants are finished products and to begin with them is to start at the end of the long story of plant life. Flowers, fruits, animal-pollination, grazing, uprooting, and cultivating are late events in the interplay between plants and animals and plants and men. For a long while during the earth's history plants were fashioning themselves and building into their environment with no concern for animal life. They were experimenting in spreading their sunlit cells with varying rigidity through the waters and over the land, and their deal with animals, which led to our civilization, was not begun until the manner of vegetation on land had been approved in the struggle for existence. To understand plants therefore it is necessary to think right back into their most unfamiliar beginnings and, because there are no

I

fossils to prompt, the enquiry into the nature of plants must be opened with a searching question.

Seaweeds and the green plants of the land are called photosynthetic because they grow from water, mineral salts, carbon dioxide, and oxygen by means of the chlorophyll in their tissues and the sunlight that it absorbs. It is this ability to live on the inorganic, or purely physical, environment that has enabled them to spread so widely wherever there is the combination of water, not permanently frozen, and sunlight. It is this ability to vitalize the physical environment that makes them the mainstay of animal life, whether directly or indirectly through a series of other animals; as our nomadic ancestors discovered, 'all flesh is grass'. Wherever there are animals therefore there must be a supply of plant matter for their existence. Now animals are much more easily understood than plants. In propelling themselves about they have more to do, and whatever they do becomes reflected in a special construction. From these special constructions of their bodies it is learnt that land animals are adaptations from sea ancestors. This is a cardinal point of zoology that must be reflected in botany. The question to be asked, then, is what are the plants on which sea animals feed?

Around the coasts seaweeds may provide the fodder but they cannot initiate the food-chains in the great oceans. There animals pursue each other with tireless brutality, the larger the smaller, and the smaller the smallest, but what can the smallest of all do? This poser has lead to the discovery of an immeasurable abundance of very simple microscopic plants in the upper sunlit waters of the sea. Although they are invisible to the naked eye, they can amass into a kind of haze, thickening here and there into green, yellow, brown, and red patches that extend for acres and are drawn out into long streaks to the horizon by wind and ocean currents. The smaller and themselves microscopic animals feed on these plants by filtering them out with bristles and sieves into a thick soup, and on the multitude of animalcules the larger animals depend; shoals of shrimps which have cropped the waters fill the whales. The plants make from the sea water and sunlight the sugars, proteins, oils, and vitamins which the animals select and concentrate in their bodies, and the tastier become the sea foods of restaurants.

Plants so minute have no ordinary names. Indeed, botany had to borrow from zoology to speak about them. Animals that float, rather than swim, were given the collective name of plankton by V. Hensen in 1887. The plants were called plant plankton (phytoplankton) or, more precisely, marine plant plankton because they have a counterpart in fresh water. Plankton is collected by towing small nets of very fine mesh, which strain off the minute organisms, or the sea water is whirled in a centrifuge until the organisms sediment at the end of the centrifuge tube. There are now many refinements of these methods, even to the machine that gathers a continuous record from the sea as the ship tows the collecting box during its voyage [1]. Many hundreds of kinds of minute plants have thus been found and their classification has become very detailed according to their microscopic characters. In general they are single cells, and the sea meadows of which the oceanographer now writes consist not of blades of grass but of separate cells, corresponding with those that build the blade of grass. They are not disintegrated plants, but plants too primitive for integration.

This is an important discovery. Nowhere else is there such a vast environment distinguished by the uniform simplicity of its plants. The shore, the land, and the fresh water contain mixtures of advanced many-celled plants and the single-celled, for they are composite environments. Only in the sea is there this major distinction, which holds the simplest in its upper waters. It cannot be said categorically that plant life began here because it is a long way chemically and physically from the conceivable beginnings of living matter to the organized cell, but it must have been in the surface waters of the sea that the plant-cell evolved, that plants took their first shape, and that plants diverged from animals. In this light a pollen-grain, which is a single cell, must be construed, and even an orchid carries a recollection of its marine ancestry. The first who clearly enunciated this starting-point of botany was A. H. Church in his essay (1919) on the autotrophic flagellate, meaning thereby the self-feeding single-celled organism that combines in a very simple way the faculties of both plant and animal.

The plant-cell of the ocean has become the unit of construction for higher plants. Whether in pine, fern, moss, or daisy, it is essentially the same. It needs careful study to see how it is organized and why, unlike the animal-cell, it should have come to a standstill in the ocean. It is a minute capsule, transparent except for one or more coloured bodies inside (figure 1). The interior contains watery spaces called vacuoles, separated by membranes of firmer substance called cytoplasm that passes from the inside of the thin wall, where the

3

cytoplasm makes a lining, across the centre to other parts of the cell. In the cytoplasm are the green, yellow, or brown bodies, called chromoplasts, which colour the cell. Then, under special optical conditions such as phase-contrast illumination, or by staining the cell content with special dyes which selectively colour different parts

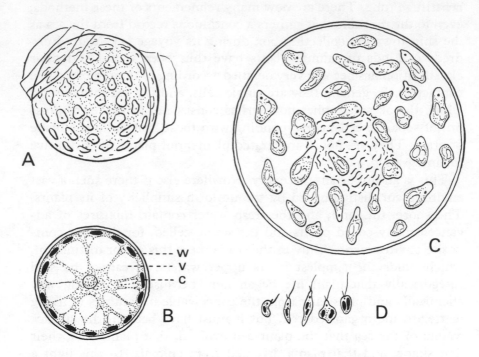

Figure 1. *Halosphaera*. A, Cell with yellow–green, angular chloroplasts, and the remains of its half-walls as transparent caps on the sides; ×400. B, Cell in optical section, showing the central nucleus, the vacuoles (v), the chloroplasts (c, in black), the cytoplasm (dotted), and the wall (w); ×400. C, Full-grown cell with the contents dividing into zoospores, each with a red eye-spot, and two chloroplasts, the centre of the parent cell containing the residual cytoplasm; ×400. D, zoospores with unequal flagella. (After Ostenfeld and Dangeard.)

of the cell, a round body called the nucleus can be made out, and this may have a central speck called the nucleolus. There is no opening in the wall and there is no free air. Thus the protoplasm, which is the name given to all the cell contents, is shut off from the surrounding sea water. Large cells may reach 1 and even 2 mm. in width but most are much smaller and have to be measured in microns (1 micron, or 1μ, being a thousandth of a millimetre). Common sizes are 10–50μ

wide, and the smallest are about 1μ wide, which means that a thousand of them could be set side by side in the space of a millimetre.

The protoplasm is the living material, but it cannot be said that one part of it is more alive than another. If the parts are separated or taken out, as can be done with skill under the microscope, they die as does the remainder of the cell. In the intact cell the parts grow and reproduce themselves by enlarging and dividing or, as it is called, by replicating and forming another image of themselves. If the cell is crushed it also dies. Dying protoplasm loses its transparency and becomes granular and opaque; its continuous structure breaks down into particles which may eventually shrink into a lump as water is withdrawn from them. By studying the manner of its dying, a great deal has been learnt about the structure of protoplasm and the parts of which it is made, but, like Humpty-Dumpty, it has never been put together again. The word protoplasm therefore, though it covers many things, is needed as an aggregate to describe the living material. For all its apparent simplicity it is the most complicated state of matter, as if it were one vast molecule that could be shattered by a microscopic blow.

The life of the plant-cell depends on five activities, namely photosynthesis, respiration, protein formation, water intake, and excretion. Collectively they are called the metabolism of the cell. Photosynthesis, protein formation, and water intake build up the protoplasm of the cell and are called anabolic activities in contrast with respiration and excretion, which break down the protoplasm and are called katabolic. Anabolism therefore must exceed katabolism if the cell is to grow. Anabolism in the plant means daylight because photosynthesis requires sunlight. In the plant-cell therefore there is an inevitable rhythm determined by the sequence of day and night, and the cell survives during the night by the excess of photosynthesis in the day.

The coloured bodies in the cell are the seat of photosynthesis [2]. In the more familiar land plants they are green from the preponderance of chlorophyll, for which reason they are called chloroplasts, but two other pigments are always present in varying abundance, namely the orange–yellow carotin and the yellow xanthophyll; these two pigments cause the yellow colour of those parts of variegated leaves unable through some deficiency to make chlorophyll. In many sea plants, however, there are other assistant pigments which render

5

the chloroplasts yellowish-brown, deep brown, pink, red, or purple and in these cases the chloroplasts are called chromoplasts. The most essential ingredient in all is chlorophyll. Like green glass it absorbs the red, orange, and blue rays of sunlight that enter it and transmits the green; but instead of merely heating up like coloured glass the chlorophyll turns the light-energy into chemical energy and causes thereby the ever-present water and the carbon dioxide dissolved in it to combine and form sugar and oxygen. The chemistry of the process is extremely intricate but the result is simple. In daylight the plant-cell removes carbon dioxide from the water around it, replaces it with oxygen, which passes out of the cell and may form minute bubbles on the outside of the wall, while sugar accumulates in the cell for the growth of the protoplasm. The sweetness of the sugar-cane is merely the exaggeration of the microscopic sweetness of every plant-cell.

Chloroplasts and chromoplasts take various shapes typical of different kinds of plants. They may be star-shaped, cup-shaped, flanged, beaded, lobed, or spread in a network over the inner face of the cell. The commonest form, and that which occurs in nearly all land plants, is a disc $1-3\mu$ wide and, simple though this appears, the electron-microscope reveals an extremely complicated internal structure. It suggests a microscopic accumulator or condenser, in the interstices of which are massed the molecules of chlorophyll, themselves as patterns of carbon, hydrogen, oxygen, and nitrogen atoms concentred on one of magnesium. Photosynthesis works by the finest interpenetration of matter and energy wherein physics and chemistry lose their distinction. Protoplasm, so elusively transparent, is organized in the finest detail, and its internal processes take place in minute quantities controlled by the minute structure, very different from the pouring of chemicals in bulk into a test-tube.

Sugar is needed for the substance and growth of protoplasm but it is not, of course, alive. It consists of carbon, hydrogen, and oxygen. Protoplasm contains many more elements, chief among which are nitrogen, phosphorus, sulphur, calcium, magnesium, potassium, and iron, with traces of others such as chlorine, iodine, boron, cobalt, and manganese. All these and many more occur as salts dissolved in the sea water. The cell extracts them as it requires and combines some of them with the sugar or its derivatives into the complicated substances with much larger molecules called proteins. Yet the proteins are not alive. They are worked into states more manifold, which, as the combination of cytoplasm and nucleus, become the living protoplasm. Thus from the inorganic matter dissolved in sea water the plant-cell under the motivation of sunlight builds protoplasm. It does not originate it but adds to it or continually replicates it on the pattern which has been handed down to it by its countless forebears; and how they obtained it is the problem of the origin of life. Plants *do* toil. They spin the fabric of living matter which animals, unable to make sugar and therefore without the loom of protoplasm, must devour. If the plants disappeared the animals would starve, but nature has many checks to over-eating.

During the day the plant-cell consumes carbon dioxide and gives out oxygen. The weaker the light the less it consumes and the less are the amounts of sugar and oxygen formed. At night photosynthesis stops and the plant-cell is then found to give out carbon dioxide by respiration and to consume oxygen just as the animal-cell does. Respiration is the oxidation, or microscopic burning, of sugar. Under the fine control of protoplasmic structure it releases in minute quantities the transformed energy of the sunlight impounded in the making of the sugar. In this sense, respiration is the reverse of photosynthesis, but the chemistry does not involve chlorophyll or light and follows another course from the synthesis of sugar. Whereas photosynthesis makes the sugar, respiration unmakes it and thereby releases energy, which is used by the cell for shifting other chemicals about for further growth. These shifts go on during the day as well as the night, but respiration in plant-cells proceeds generally at a much slower rate than photosynthesis, which uses up immediately the carbon dioxide formed by respiration in the day; it is said that photosynthesis masks respiration in the daytime.

Animals certainly respire more rapidly than plants and, weight for weight, they generally respire more carbon dioxide in a given time than plants photosynthesize oxygen. But plants are vastly more numerous than animals; the mass of plant protoplasm on the earth greatly exceeds that of animal. The comparison of respiration, in which animal life revels, with photosynthesis brings out a fact that we are apt now to overlook with artificial lighting. The one surpassing process on this earth that makes life possible is photosynthesis, and it employs the sun's energy which comes to the earth; the energy of life is not earthly. No wonder therefore that in the sea, which receives most of this sunlight, there is the world of simplest plants suggestive of the primitive origin of plants.

7

The cell-wall restricts the protoplasm. It prevents it oozing out and it prevents harmful creatures such as bacteria from coming into direct contact with it. The cell-wall is also a stabilizer. Protoplasm is eight- to nine-tenths water, some parts being far more aqueous than others; nevertheless, it is more concentrated than sea water, which, through the physical process of osmosis, continually enters the cell. The protoplasm swells, or hydrates, and would burst if it were not for the strong retaining wall on which it presses to the limit of its elasticity. So the cell-wall maintains equilibrium at maximum capacity with water continually diffusing in and out. This is the normal turgid state of plant protoplasm in the sea and it is the state that plants have to maintain on land for their normal working; non-turgid plants are wilted plants and they work poorly if at all.

As the protoplasm increases its substance by photosynthesis and protein formation, it also presses on the cell-wall. The wall, though rigid to water pressure, yields to growing pressure and increases its surface by adding more wall material between its pre-existing parts. It is like a globe of glass-fibre that grows by inserting more fibres between those already in position. At the same time, more fibres are added within the outer layer and, as the wall thickens, it loses this extraordinary power of extensibility. In time therefore the wall limits the size to which the protoplasm can increase. The cell may then divide into two, when each half will grow a new half-wall or the protoplasm will come out of the cell to start again. In this case the wall is left behind as a dead shell; for dead it is like any other part of the cell that is not the protoplasmic whole.

In three profound ways the plant-cell differs from the animal-cell. The chromoplasts, which photosynthesize, are lacking in the animal, which is the reason why it cannot subsist on inorganic matter and its food must ultimately be vegetable. Theoretically there could have been a world of plants only, but protoplasm, being so complex, may break down and lose part of its equipment; if so impaired it can still find a means of survival, then a new state of living matter may be realized. By loss of chromoplasts the animal came into the sea and the fungus on to the land, both to discover the inexhaustible issues of chlorophyllous plants. Secondly, the firm cell-wall prevents the plant-cell from eating in the animal manner; particles of food cannot be taken through the wall. Thirdly, the wall prevents the plant-cell from quick changes in shape; musculature is denied the plant, whose protoplasm needs the stiff surround to maintain its working. Plant protoplasm is cast in moulds; the animal shapes itself. What appears to be a hindrance, because it makes the organism stationary, becomes nevertheless the mainstay of plant evolution; the substance, tenacity, and rigidity of the cell-wall make the framework or skeleton of the higher plant. Every feature that we attribute to the form, shape, texture, and presence of a higher or many-celled plant depends upon its cell-walls. A leaf, petal, or fruit is shaped by its cell-walls; trees are the expression of cell-walls; timbers differ according to their cell-walls; toadstool and seaweed, inflorescence and root, are the outcome of pushing and pulling cell-walls. Botany, in fact, is founded not on plant protoplasm, which is yet too elusive, but on cell-walls made by the internal protoplasm in carrying into effect its development, its evolution, and its world exploration. The skeleton of the plant is a casting round every cell, far more minute and expressive than the skeleton, internal or external, of the animal. In botany, strange to say, we study largely the dead casts to discover the living.

How does plant protoplasm make a solid wall, pervious to sea water, out of sea water? This wall is not a chemical precipitation where protoplasm contacts the water and it is not a special arrangement of protoplasm itself on its surface. It is a tissue of very minute fibres of cellulose, arranged in layers with the fibres crossing in different layers so that a texture is built up like the fabric of a rubber tire. Cellulose consists of sugar molecules strung together by the removal of molecules of water. Thus by taking water chemically out of the sugar, which it makes abundantly in the day, the plant protoplasm spins minute lengths of minute thread that are bundled together and the bundles are placed at the surface of the protoplasm and compacted into the wall. None of this can be seen under the ordinary microscope, which shows the wall as homogeneous glass or faintly striated by optical interference. It is the development of the X-ray analysis of cell-walls and of electron-microscopy, which, with more penetrating powers and greater magnification, have revealed the finer structure [3]. Measurements of cellulose fibrils have now to be made in angstrom units of a ten-thousandth of a micron: the bundles of cellulose fibres, or microfibrils as they are called, measure 50–100 angstrom units in width. The same pattern has been found in a fair sample of the plant kingdom, and this generality confirms the impression that cellulose, and some other and similar substances, in the cell-wall are excreted from the plant protoplasm as the excess of sugar in the excess of photosynthesis not needed for making

9

protoplasm; their very inertness, chemically, makes them suitable for building, and building is plant evolution.

The nucleus is the chief body of the cell. All living cells, whether plant, animal, or fungus, contain a nucleus, and yet if separated from its cytoplasm the nucleus, just as the chromoplasts and the cytoplasm itself, also dies. The nucleus contains the hereditary material that directs the activities of the protoplasm. It stores, that is, the memories of how to grow, react, and propagate, and one remarkable point is that the nuclei of single-celled organisms, such as marine planktonic cells which have such simple lives, appear very similar to those of the highest, whether in a wheat plant or a human body; primitive nuclei clearly had the capacity for evolution, much as the brain of primitive man had the capacity for education. The nucleus governs cell chemistry by means of enzymes, which are protein bodies that cause particular chemical reactions leading to the building up and the breaking down of cell ingredients such as the sugars, oils, proteins, vitamins, cellulose, and other organic substances. Enzymes may be served out by the nucleus or delivered by the cytoplasm on receipt of a message from the nucleus. An extremely thin membrane surrounds the nucleus and separates it from the cytoplasm. On consideration of what happens on the two sides of the membrane, the one ruling and the other obeying, the membrane appears as one of the most remarkable refinements in the whole world.

The nucleus has also a detailed structure, in which the chromosomes are conspicuous. These bodies, so-called from the way in which they can be coloured by certain dyes, appear as minute rods or spheres at the time when the cell divides. Their number and shape is constant for every kind of plant and animal. The numbers in plants vary from three to more than five hundred and they are presented in the nuclei of the same organism in two ways: either as x chromosomes or as $2x$ chromosomes, that is, for example, as three or six or two hundred and fifty or five hundred. These states are called respectively the haploid and the diploid and, as will be explained, they are connected with sexual reproduction, which doubles the number, and with the complementary process of halving the chromosome number, which is called meiosis. Neither the number nor the obvious size of the chromosomes is a measure of their ability to produce an organism, for this seems to depend on the ultramicroscopic chemical structure of the substance deoxyribonucleic acid

(DNA) of which they are chiefly made. So, contrary to common sense, as happens so often in matters of ultra-structure, lower plants with simple bodies and simple behaviour may have more and larger chromosomes than higher plants possessed of so much greater complexity. The molecular organization of heredity and of nuclear activity is a modern subject, rapidly expanding, yet based on a very limited number of plants and animals, and it cannot be fitted yet into a coherent account of plant life in general; an analogy must suffice. The plant-cell, so astoundingly transparent, minute, and apparently simple, has at the molecular level a complexity comparable with the goings on of a city. There is the inner city and the government within the nuclear membrane, outside of which lie the streets, houses, shops, factories, post-offices, police-stations, hospitals, lighting and sewerage systems retained by the outer wall; and the lesson is that everything living is built throughout in minutest detail.

The plant-cells photosynthesize and grow during the day. When they have reached a certain size, typical for each species, they divide into two. This form of multiplication is called binary fission, in contrast with multiple fission when the cell contents divide into four or more smaller units. Binary fission is the ordinary means of multiplying the numbers of individuals in the plant plankton, and it seems generally to occur at night. Nocturnal cell-division has been observed, indeed, throughout the plant kingdom and may be a persistent heritage from planktonic life and a peculiarity of the revolving earth. Though some plants are said to perform sleep-movements, such as the folding together of clover leaflets at dusk, and seeds are said to have periods of dormancy before they sprout, plants do not sleep in the manner of animals. After the photosynthetic surfeit of the day they toil internally and most of their growth may be performed at night.

The nucleus is the first part of the cell to divide. First, its membrane disappears. Then the chromosomes contract, duplicate themselves, and become arranged in two complete sets that separate and form each a new membrane (figure 2). When the two daughter nuclei are formed the chromosomes expand into their active state and the cytoplasm is divided into two parts by a fine membrane that forms across the cell between the nuclei. On either side of the membrane a new cell-wall is formed, continuous with the previous half-wall, and lastly the cells separate and begin their individual lives. There are many variations in detail of this process but in broad outline it is the

standard method of cell-division evolved in the single-celled planktonic state, and inherited by all higher plants. The misfortune is that
we know more about cell-division in the familiar higher plants than
we do of its exciting primitiveness in planktonic cells.

The chromoplasts divide or duplicate themselves independently
of cell-division and they keep up their numbers in their own time.
The daughter-cells grow to their full size without further nuclear
division or increase of nuclear size, but the chromoplasts increase
their numbers, if not their size, to fit the enlarging surface of the cell.

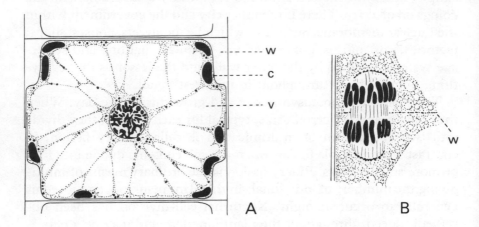

Figure 2. Diatom *Melosira*. A, Cell with central nucleus, brown chromoplasts
(c, in black), vacuoles (v) separated by strands of cytoplasm containing small
granules, and the siliceous diatom shell in two halves (w); × 700. B, Nucleus,
now on one side of the cell, dividing with the chromosomes doubled into two
sets, one for each daughter-cell, and the new cell-wall (w) beginning to form;
× 2000. (After Cholnoky, 1933.)

There is evidence to suggest that even vacuoles, which seem mere
watery drops in the cytoplasm, are self-propagating bodies and that
they may contract into specks, duplicate, and enlarge again. There
are also other minute bodies in the cell called mitochondria, concerned with the formation of enzymes, which maintain themselves
by independent duplication; in minute structure they resemble the
chloroplasts but have no chlorophyll.

If cell-division occurs once every twenty-hour hours, the number
of cells will increase a hundredfold in a week and rapidly become
astronomical. So it is that, under favourable conditions of sunlight,

temperature, and mineral content of the water, the normally invisible suspension of plant-plankton thickens into a haze and into coloured patches. The numbers can be estimated by counting a proportion of the cells sieved or centrifuged from a definite volume of sea water. They vary from one cell per cubic millimetre to two thousand for minute cells less than 10μ in width. In coloured patches, cells of $20-50\mu$ in width may average fifty per cubic millimetre and they may extend in this quantity to depths of 80 m. [4]. By sampling throughout the year it has been estimated that in the English Channel 1400 metric tons of plant plankton are produced per square kilometre per annum; figures for hay-crops (weight wet) in England work out in comparison at 1800 metric tons per square kilometre per annum.

Plant plankton in the upper illuminated water of the sea is as important to animal life as the veneer of vegetation of comparable thickness on the land. It removes the carbon dioxide and oxygenates the water; indeed, because of the solubility of carbon dioxide, the sea is the great stabilizer of the amount of carbon dioxide in the atmosphere. The plant-cells remove the mineral matter dissolved in the sea, particularly the nitrates and the phosphates, and turn it into protoplasm in a form that animals can assimilate. So avid are the plant-cells that it is even considered that this essential mineral matter is present in the surface of the sea more in the form of plant and animal bodies than as dissolved salts. The great outbursts of plankton generally last only a few weeks. The geometrical increase in cell numbers may well deplete the supply of mineral matter and so cause its own decline, but others consider that it is the increase in animal life, feeding and multiplying on the plant plankton, that devours it. The abundance of nitrate and phosphate in sea water is, however, the main limiting factor to the richness of the sea pastures.

There are places nevertheless where this mineral matter is renewed. Rivers bring down silt and dissolved salts, and the shallow seas into which they discharge are continually churned up by water movement. The North Sea, the Baltic, the Yellow Sea, the Red Sea, the Adriatic, parts of the Caspian Sea and the waters of marine gulfs have their annual rhythms of planktonic increase, partly seasonal in relation to temperature, sunshine, and storms, and partly in tune with the rivers. In contrast, siltless coral seas and mid-oceans appear as plant deserts, but the blueness of their waters means rather the lack of renewal of the mineral matter than an actual rarity of plant plankton. In its ubiquity this plankton maps out the subtle creep of the ocean itself. Where deep waters, rich in mineral matter accumulated in cold, dark, sinking, and plantless currents, upwell towards the poles, the plant plankton multiplies and becomes the fodder of the shrimps of the whaling seas. Where cold currents drift equatorwards along the shelves of continents, such as the Humboldt current off Chile and Peru, the Benguela current off South-West Africa, the Labrador current, and the Kamchatka current off Japan; where submarine ridges cause slow vortices of deeper water to ascend, as off the Grand Banks of Newfoundland, and the coasts of Iceland, the Faeroes, and West Scotland; or where alternating monsoons, driving the surface waters before them, stir up the counter currents in their depths, as in the Arabian Sea; there plant plankton thrives, fish shoal, sea birds crowd, and fishing fleets collect. The sea water circulates. Deep water replaces that at the surface which has been depleted of the materials needed for life.

Sunlight in passing through sea water is gradually absorbed. The order of absorption of colours follows the order of the rainbow, beginning with the red. At 30 fathoms a bluish light prevails, below which is increasing darkness, except for the phosphorescence of deep sea animals. In clear tropical waters with the sun overhead light sufficient to fog a photographic plate may reach 1500 fathoms. In the recent bathyscaphe exploration into the Pacific deep of 6000 fathoms off the Marianas Islands, sunlight became apparent to the dark-adapted human eye, during the ascent, at 200 fathoms below the surface. But such light, as well as the phosphorescence, is too feeble to maintain plant life, for which 150 fathoms seems to be the maximum, and 50 fathoms the more usual limit. In coastal waters, where the light is absorbed or scattered by the silt in suspension, 10 or 20 fathoms may be the limit of effective photosynthesis. As the average depth of the sea is estimated at 2000 fathoms, it is the merest skin in which the plant plankton lives.

Consider then this plankton. It drifts in currents broad and narrow, fast and slow, deep and shallow. It warms up and cools down in different latitudes, and the changes of temperature inhibit some kinds of plant-cells and promote others. Cooling surface water at night sinks and carries plankton down while warm water ascends and carries it up. It plunges in the wake of fish. It blows off in the spray of white horses, to shrivel in the wind. Heavy rain dilutes the surface and dry weather may concentrate it; changes in the concentration of the water may kill the plant-cells by disturbing their internal balance.

Anything that alters the temperature and saltness of the water affects its density and sets up local and regional currents with counter currents and vortices. By the twenty-four hours and by the seasons there is a continual mixing, far and near, in the skin of the ocean, and a flowing of the plankton.

The plant-cells are normally buoyant. The oxygen formed in photosynthesis may float the cells in the day, as it floats the green scum of ponds, and the oil in the cells, which is derived from the sugar of photosynthesis, is lighter than water. But what will happen if currents carry the cells into the limitless darkness? The cells will use up their reserves of oil and sugar through respiration until they die. Dead plant plankton always sinks. A light rain of the dead and dying plant-cells must fall into the depths. Some may reach the bottom, where their remains will be decayed by sea bacteria, but most are eaten at some stage or other by minute animals and their larvae. Then these are eaten by other animals, which are eaten in their turn, while the largest are dismembered in their infirmity; the rain of plankton gives place to the droppings of animals, which have lived distantly, if not immediately, on the plant-cells at the surface. Protoplasm descends in stages and is finally decomposed at the bottom. Here the mineral matter is freed by bacterial action and re-dissolved under the abyssal conditions of great pressure, deficiency of oxygen, and excess of carbon dioxide. The deep waters are enriched from the upper, and their upwelling becomes the fertile surface. Whatever the original nature of the sea it is now, after the lapse of geological time, diversified by the activity of primitive plant life into a photosynthetic or anabolic skin and a decomposing or katabolic interior.

Chapter 2

Life-cycles

THE ROUND, yellowish green cells of *Halosphaera* (figure 1) are common in the surface waters of warm seas. Their colour is caused by the rather high proportion of yellow xanthophyll, which accompanies the chlorophyll in numerous, small, and rather angular chromoplasts. The rest of the cell consists of large vacuoles separated by membranes of cytoplasm, the nucleus, and the retaining wall. The cell grows considerably during its life without dividing, and can reach a width of half a millimetre. At times, to accommodate the increasing protoplasm, the wall splits and it is then seen to consist of two halves. Two new and larger halves form inside the older, which may remain on the sides of the cells as little domes of organic glass. The cellulose wall is stiffened as it ages by the deposit of glass (silica) in it and this causes the loss of extensibility. Silicification of the cell-wall is a widespread, if unexpected, property of plants. It is found in various kinds of planktonic cells, especially the diatoms, and among many land plants such as the horsetails (*Equisetum*) and the grasses, but the chemistry of this excretion is little understood.

Binary fission has not been observed in *Halosphaera*. Instead there occurs multiple fission. The nucleus of the cell divides into two. These two also divide and so do their four daughter-nuclei. The process continues until a large number of small nuclei are formed. Then around each nucleus the cytoplasm collects to form minute bodies of protoplasm about 10μ wide, and these new plants, for such they are, escape out of the cell as the halves of the wall gape, and swim off on microscopic journeys. The new organisms are called zoospores because they are spores that multiply the plant and they move actively like animals. The zoospore has a nucleus, a few chromoplasts in the cytoplasm, an extremely thin outer membrane but no cellulose wall, and a tiny red spot near the end which carries two fine unequal hairs. Each hair is called a flagellum; by lashing the flagella the zoospore swims. The red spot is believed to be a very simple eye. In other single-celled plants it has been shown by experiment under the microscope that the end with the red spot can detect light, to which flagellar movement is directed. The eye-spot may even have a small lens.

The zoospore begins to grow as soon as it has escaped from the parent cell. It is a new thing with its own liberty, but heredity clamps down. It begins to deposit a cell-wall; it loses its flagella and eye-spot, perhaps reabsorbing them; it turns into a plant globule, which, by periodic formation of a larger wall, grows to the adult size, fifty times the original width; it then divides itself into zoospores and passes the hereditary custom on to the next generation.

How long this all takes and whether it is all so straightforward is undetermined. Such is the condition of most of our meagre knowledge about these simple plants. It is difficult to study their lives. Microscopic work is extremely tiresome on a rolling ship, and it is difficult to keep the planktonic cells active in an aquarium on land. What can be related is mainly compounded from the observations of cells in different stages of growth. It has not been possible to watch a single individual throughout its normal life, as one may watch a seed or indeed a land spore become a flowering plant or a fern.

This unexpected behaviour of *Halosphaera* is significant. It implies that the simple plant-cell is not the beginning of plant-life or of botany, but the result of growth from a smaller mobile organism without a cell-wall. Many such organisms are known. They are called flagellates and, because of their small size, generally 2–10μ wide, they form what is called the dwarf plankton (nannoplankton) generally to be studied by centrifuging the sea water. Having chloroplasts, or chromoplasts, they photosynthesize as plants. Many have no firm cellulose wall and their bodies therefore can change shape by contraction, as animal cells do. The flagella, conferring motion, may be thought also to indicate an animal nature, but flagella occur in the reproductive cells of most lower plants and, indeed, in a few higher plants such as the cycads and the maiden-hair tree (*Ginkgo*); they are not a criterion of zoological status. Other kinds of these flagellates have the faculty of forming cell-walls, which mark them out as truly botanical. Yet others have few or no chloroplasts and, being without a cellulose wall, they can swallow or engulf food particles, such as bacteria or other flagellates, and digest them in their cytoplasm in the manner of *Amoeba*. Such flagellates are animal rather than plant.

17

The flagellates, in fact, as a whole combine the elementary features of plant and animal. Some are very simple motile plants and are called plant flagellates (phytoflagellates); others are non-photosynthesizing animal flagellates (zooflagellates); others are half-way between the two. They form in their simplicity, minuteness, and generality the pedestal of biology on which its twin subjects, botany and zoology, stand [5]. Thus both sciences claim the photosynthetic flagellate *Euglena* (figure 3), which reproduces by binary fission and, being a zoospore itself, does not undergo multiple fission into zoospores as the plant-cell *Halosphaera*. This consideration leads to the second point.

The large non-motile cell of *Halosphaera* is its adult, vegetative state. The zoospore, from which it grows, is its young, larval, or flagellate state. Other plant plankton, such as diatoms – to be described later in this chapter – behave in the same way. Here is the beginning of development from a simple, motile, and naked reproductive state to a larger, non-motile, adult or vegetative state enclosed in a cell-wall, similar to the development of the adult in a higher plant or animal, but much simpler. The theory is that such development represents what has been the history, or past evolution, of the organism. *Halosphaera* is so simple that there can hardly be objection to the conclusion that small plant flagellates preceded in time the presence of plant-cells (non-motile, possessed of a cell-wall) in the sea. When higher plants and animals are considered in this respect, however, many complications enter into their development, both short-cuts and elaborations, so that their life-histories are not strictly repetitions of their evolution. The old biological saying, often now discredited, that an organism in its development climbs up its own evolutionary tree is both generally correct and commonly erroneous. Nevertheless, the theory is – and the lives of the lower many-celled plants such as seaweeds, mosses, and ferns bear it out – that plants began as plant flagellates that evolved the non-motile, cellulose-walled, plant-cell such as reigns in the marine plankton, and then proceeded to build the many-celled plant body under the very different conditions of the seashore. On this turns indeed the modern science and classification of seaweeds. In order to reproduce they revert to zoospores, which can be classified and set in the scheme of biology alongside the planktonic flagellates. To regenerate, that is, many-celled plants must go back to a flagellate start in the same way as *Halosphaera*, and out of this their chromosomes recollect, according to their kind, the manner of development that ends with the specific adult being. The photosynthetic flagellate therefore is the

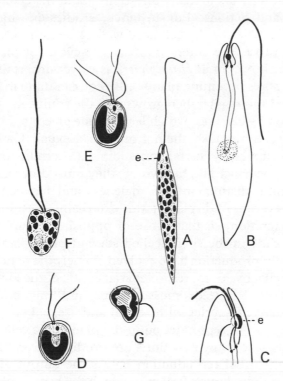

Figure 3. Marine plant flagellates, with the chloroplasts shown in black; ×900 (B and C more highly magnified). A, *Euglena*, with red eye-spot (e) and without cell-wall. B, *Euglena*, showing the two strands connecting the forked base of the flagellum with the nucleus (after Johnson, 1957). C, *Euglena*, the front end showing the connexion of the eye-spot (e) with the base of the flagellum (after Chadefaud, 1960). D, *Chlamydomonas*, with a large basin-shaped chloroplast, eye-spot, two flagella, and a cell-wall. E, *Carteria*, as *Chlamydomonas*, but with four flagella. F, *Chloramoeba*, with many small yellow–brown chromoplasts and two unequal flagella, without a cell-wall. G, *Nephroselmis*, with one lobed yellow–brown chromoplast and two unequal, laterally placed, flagella, without a cell-wall. (A, D–G, after Schiller, 1925.)

simplest state of plant life. Its circumstances are the sunlit sea where it has proceeded a step further to evolve the non-motile plant-cell, and thereby to introduce the life-cycle with adult and reproductive stages.

19

Knowledge of marine flagellates is very fragmentary. Their diversity, however, offers a great field of exploration wherein will surely be found the answers to many questions about the evolution of basic structures and actions. For instance, studies in this field must eventually reveal how the cellulose wall came into existence.

Another procedure enters into the life-cycles of plant plankton, but whether it occurs in *Halosphaera* is as yet uncertain. Instead of all the zoospores becoming plant-cells they can unite in pairs and the union, called the zygote, then grows into the adult cell. This is sexual reproduction by gametes, which is the name given to these fertilizing zoospores to distinguish them from the asexual. Gametes do not unite indiscriminately; certain ones join with certain others. Though similar in appearance and behaviour, they must differ internally with some chemical distinction into maleness and femaleness and they must have a means of displaying this externally, presumably by some chemical excretion, so that those of opposite sex are attracted together. Already in the single-celled state of plant plankton the two processes of reproduction have evolved, namely the sexual by gametes and the asexual by zoospores, so characteristic of the lower half of the plant kingdom. Seaweeds, mosses, ferns, and fungi embark thereby on extravagant life-cycles with sexual and asexual stages.

Both methods of reproduction in the planktonic cells cause multiplication because many new units are produced from the one adult, even though sexual reproduction halves the output. They are two more ways besides binary fission in which the haze of plant life in the water may thicken. All methods depend on satisfactory growth to make enough material for the new units. In many-celled plants and animals the many-celled body ages and dies after reproduction. Plant flagellates in contrast seem not to age, except so far as time is needed for multiplication, and they have no dead body. *Halosphaera*, as a plant-cell, both ages as it increases in size and shows the beginning of death in discarding the old cell-wall from which the zoospores escape. We do not know the reasons why planktonic cells should merely undergo binary fission or why they should modify this into multiple fission, or why multiple fission should lead either to sexual or to asexual reproduction.

When two gametes conjugate the cytoplasm of both intermingles and the two nuclei combine into one, but the chromosomes of each gamete remain distinct. The zygote therefore has within its now diploid nucleus twice as many chromosomes as either gamete. Since chromosome number is constant within a species, somewhere in the life-history the number must be halved and haploid nuclei re-formed. In higher plants it is known that this halving, or 'reduction' as it is called, of the chromosome number occurs at a special nuclear division. As it involves steps different from ordinary nuclear division and re-establishes the haploid state, this special division is given the name of meiosis; ordinary nuclear division, during which the chromosomes duplicate and separate to form two nuclei without a change in chromosome number, is called mitosis. The binary fission of plankton cells and the cell-divisions that lead to the many-celled bodies of higher plants are mitotic, though the nucleus may be haploid or diploid. Normally, only conjugation turns the haploid into the diploid, and only meiosis restores the haploid. These two events are therefore a unique couple which plays an increasingly important part in the lives of plants as they become more highly evolved. The couple gives an outlet for successful variation, very imperfectly understood, in the attached seaweeds, unlike the strict discipline of most animals in whose diploid bodies meiosis occurs only at the formation of the gametes.

For reasons concerning the detailed pairing of the chromosomes, which are important in genetical theory, meiosis is always followed immediately by a second and more or less mitotic division to give four haploid nuclei. There may then be several more divisions to give eight, sixteen, thirty-two, sixty-four, or 128 haploid nuclei and as many protoplasmic units. Meiosis is therefore an antecedent to multiple fission, for which reason it is suspected that the zoospores of *Halosphaera* may in fact be haploid gametes.

Since the lives and indeed sometimes the construction of higher plants cannot be followed without a knowledge of the haploid and diploid states of the nucleus, meiosis, and conjugation, a simile may help. A chromosome is an entity with an ability; it exists with a structure and function, and it duplicates itself. Other chromosomes in the same nucleus are other entities. The set of chromosomes may be likened to a football team of eleven players or entities with a special part to play in the game. During mitosis two complete teams are formed and their manner of play is thus multiplied. During conjugation two different teams meet. During meiosis these two teams separate, but in a peculiar way such that, before parting, every player pairs with his counterpart, outside-left with outside-left, right-back with right-back, and so on. Then, in parting, some players exchange sides; the teams are shuffled, but each player keeps his own position in the game and the two teams are complete. If, however, the two teams decide to combine into one with twenty-two players, every player with a reserve, that is the diploid or doubled state which may be propagated by mitosis until meiosis reorganizes it into two haploid or normal teams.

Plant plankton has at least two kinds of life-cycle. In the diatoms, and possibly in *Halosphaera*, the adult cell is diploid and propagates by binary fission. Then at certain times the cell undergoes multiple fission and forms many flagellates, which, it is believed, are haploid and act as gametes. The zygote then restores the diploid condition and grows into the adult cell. In other cases such as the motile green plant-cell *Chlamydomonas*, with two flagella, the adult cell is haploid. It also propagates by binary fission and, at times, it also undergoes multiple and mitotic fission to form four, eight, thirty-two, or sixty-four small naked flagellate cells which may behave either as zoospores and grow directly into the haploid adult or they may behave as gametes and conjugate. In this case the zygote undergoes meiosis and forms usually four haploid zoospores, which grow into the adult haploid cells. Thus without the study of the chromosome behaviour, which is laborious and exacting in these elusive organisms, their nature and life-histories cannot be properly understood.

Chlamydomonas shows that conjugation is followed immediately by meiosis; that is, as soon as the chromosomes, or players in the two teams, have been shuffled around they separate. The diatoms on the other hand maintain the superiority of a double team throughout most of their lives. The double team, or diploid state, comes eventually to predominate in higher plants, such as the large brown seaweeds, the ferns, and the flowering plants, just as it is the normal state in animals. Conjugation and meiosis, by shuffling the players, produce new teams, some of which will be better than others and more likely therefore to succeed in the game of life, while the poor teams are not only beaten but eliminated. Thus the sexual act of the union of gametes provides one of the means by which natural selection can improve the race of an organism: it is an elementary act of protoplasm established in the plankton stage of plant-life.

The flagellum is also a product of planktonic protoplasm. In the plant kingdom, except in the plant flagellates, it occurs only on the zoospores or the gametes and is absent from the typical non-motile plant cell. Animals cells, in contrast, without a cellulose wall, may variously retain the flagella as the ciliated surface layer of cells, whether internal or external. These microscopic hairs, whether called flagella or cilia, are constantly about 0.2μ in width and have in all cases, plant or animal, a constant basic structure. The electron-microscope has shown that they consist of eleven strands united in a sheath so that nine strands form a ring around two central ones. They are not simple threads of surface cytoplasm, such as the much finer bacterial flagella are, but complicated structures formed as if they had been thrust through a die of eleven holes at the surface of the cell [6]. Their intricacy serves to show how little as yet our coarse senses can follow the problems of planktonic life. Plant flagella draw the body of the zoospore or flagellate after them, for which reason they are described as anterior, or tractor, and are said to be attached to the anterior end of the cell, which goes first. In animal spermatozoa the flagellum is posterior and propels the body of the spermatozoon in front of it. According to this distinction, the disputed *Euglena* is a plant flagellate with anterior flagellum.

Why should planktonic flagellates transported in the turmoil of the ocean have means of microscopic movement? They may drift several miles in a day and vertical currents may carry them through several fathoms' depth. To think out possible uses of such a structure will help to discover its value to the organism, though it will not explain its structural and metabolic origin. The flagellum, once made, could have assisted in several ways. First, for movement one thinks of attack and escape, and some of the flagellates move with such speed and agility that they are only with difficulty kept in view under the microscope. In the beginning, however, plant flagellates must have preceded the animal and presumably, as they did not devour each other, this use would have been a subsequent development unless there was a predaceous sub-flagellate ancestor to prey on them. More real is the use of the flagellum, in conjunction with the red eye-spot, to carry the plant flagellate towards the light. Sinking into the darkness is the great danger, and, of course, at these depths of 50 fathoms or more, sea movement is much diminished. To be able to swim up towards the light could mean salvation. Thus may have evolved the need to perfect sensitivity to light or, as it is called in botany, phototropic response and to couple it closely with flagellar activity. In many flagellates, as *Euglena* (figure 3), it has been found that there are special strands of cytoplasm connecting the eye-spot and the flagellum, as if they were a local transmitting system.

23

Then, in comparatively still water or in a current carrying the flagellate along with it, the metabolic activity of the protoplasm will soon extract the carbon dioxide and mineral matter from its immediate surrounding. Further supplies will depend on the slow process of diffusion, but, possessed of a flagellum, the protoplast can keep moving through the water and exploit new surroundings. One can think, too, of the gametes and the impossibility of finding a mate in the sea unless the gametes are possessed of activity, which will bring them together either by leading them towards a common source, as in the direction of light, or by attracting them with chemical excretions from their protoplasm into the water. In this case a gamete must be attracted not to its own excretion but to that of its future partner; thus conjugation implies at the outset a difference in behaviour, an incipient maleness and femaleness in the conjugants. Zoospores and gametes are produced at the end of a period of growth. It may be that they are the means whereby the sinking plant-cell multiplies itself into motile forms, rejuvenates, and increases its chance of returning into the well-lit or photic upper zone of the sea. The twilight depths where light is becoming too feeble for photosynthesis may be the critical region in the environment where plant plankton evolved its special activities, over and above photosynthesis, and its life-history.

It may be thought that these problems are too difficult and too distant from plant life in general. But higher plants are built up from cells, and what properties the plants have are the consequence and sublimation of the properties of these cells. The ocean is the environment where the plant-cell indulges uniquely in uncombined existence. There presumably its properties were evolved to maintain photosynthetic life. When these properties are found in pines and oaks, for instance, in the form of cell-walls, chloroplasts, mitosis, meiosis, conjugation and a life-cycle beginning with a single cell, it must be inferred that they have been inherited from a single-celled oceanic progenitor. The properties are the planktonic legacy that has been developed into the vastly more effective lives of pine and oak by their ancestors, as they moved through circumstances more and more remote from the ocean.

The great lesson of botany begins to appear. Plants may do as much as the environment permits for photosynthetic activity. A more complex environment permits a more complex plant; complication does not occur among plants just for the sake of evolution.

On entering a more complex environment the plant carries into it, however, a legacy from the past and with this ability or limitation it proceeds. The more environments its ancestors have passed through, the greater is the plant's endowment, though the benefit may be small if the talent has not been used. Some ancestral plant flagellates improved themselves into the oceanic cell. This absorbs energetically over its whole surface and remains a selfish unit; combination with other cells into a many-celled body would hinder the absorbing power of every cell without conferring an advantage. In the next environment on the sea coast the table is turned and combination into the many-celled body becomes the leading feature of the sea-weed: on land the multiple body is stiffened into the tree. Other flagellates, however, have persisted in the ocean without talent for evolution, and some of them have been carried by estuary, lake, salt-marsh, or other aquatic route on to land, where they have remained flagellates of fresh water, drips, puddles, and bogs. Such dispersal, with local adaptation, is very different from progressive evolution when the plant develops a new mode of life under a new set of circumstances. The new mode evolved by the selection of structures and actions to meet and overcome the photosynthetic problems in the new environment; on further colonization that new mode of life becomes the legacy inherited without the originating circumstances. The clouds of pollen emitted by pines and oaks in spring for reproduction are at once understandable as aerial single-celled plankton; in a later chapter it will be shown how the history of pollen can be traced by analogy with fern spores. If there was not such guidance to botanical thought, all sorts of fantastic ideas might be developed about the need to blow pollen from one tree to another. So it is necessary for the understanding of plant life to follow it from the lowest level. In a broad way the successive environments and their problems are perceived, but the explanation of the means of plant evolution lags far behind.

Predominant in the present seas are elaborate plant-cells of three kinds, namely the diatoms, the peridineans (or dinoflagellates), and the coccolithophores [7]. In all of these the chlorophyll is accompanied by a brown or yellow–brown pigment that obscures the green. Their cell-walls are stiffened also by mineral matter, siliceous in the diatoms and peridineans but calcareous in the coccolithophores.

The stiffened wall of the diatom has the form of a box in two almost equal halves, one overlapping the other, and hence the name of

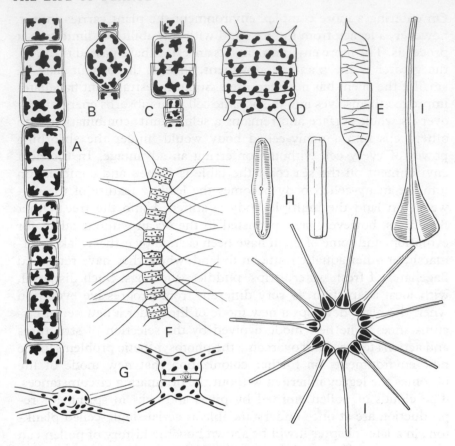

Figure 4. Marine diatoms (at different magnifications). A, *Melosira* (diagrammatically simplified), showing the manner of cell-division and the arrangement of the cells in a filament, the yellow–brown chromoplasts star-shaped; ×500. B, *Melosira*, forming an auxospore and restoring the cell size; ×500. D, *Biddulphia*; E, *Rhizosolenia*, with extra pieces of shell put into the side walls; ×250. F, *Chaetoceras*, with four fine processes of the shell (not flagella) from each cell; ×250. G, *Chaetoceras*, enlarged to show the cell form; ×500. H, *Pinnularia*, an oblong or pennate diatom; ×500. I, *Asterionella*, a pennate diatom with one end of the cell attenuate, and the cells set into a circular colony; ×650; also a single cell, ×1000. (After Hustedt, 1959.)

'two atoms'. The box may be circular, as a pill-box, or oblong, and this difference distinguishes the two main subdivisions as the centric (circular) and the pennate (oblong) diatoms. Actually, the siliceous plates forming the top and bottom of the box may be separate from the sides and new pieces can be intercalated on the sides, thus complicating the appearance. The plates, too, are generally marked or sculptured with fine lines and dots, which are actually pits or hollows. The markings are extremely diverse and descriptions of an immense number of species have been based on them. Identification from this ornamentation, as the markings are generally called, needs the very best microscopes and microscopic techniques. Slides made with standard diatom shells used to be test objects for the resolving power of microscopes, but this enthusiasm of the last century has died and diatoms enter little into general botany. Nevertheless, interest is returning. The markings of the shell display the elaborate structure of the surface cytoplasm, which the electron-microscope can investigate, and as living beings they show the complexity that plant plankton can attain. Because of their vast numbers in the sea, they are one of its main, ultimate sources of food and there is in consequence increasing research into their manner of growth and dis-

Figure 5. Diagram of cell-division in a diatom, covering four generations. Letters a–f indicate successively smaller and inner halves of the shell.

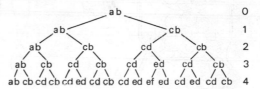

tribution. In the coastal waters of the British Isles diatoms produce a spring crop and an autumn crop on which the animals and fisheries depend: the summer crop consists chiefly of peridineans. Diatom cells measure 10–200μ wide, according to the species, but recently there have been discovered in the Gulf of Mexico what seem to be exceptionally minute species 1–2μ wide [8]. If such are common and widespread, they will be an important element in sea economy that has not been reckoned with.

When the diatom divides into two, which seems to happen usually at night and, in full growth, every night, the two halves of the box are thrust apart and each daughter-cell (produced mitotically) forms a new inner half. As one daughter has the parental inner half and then grows a new inner shell, some daughter-cells will become successively smaller (figure 5). When the lower limit of size is reached, according to the species, the protoplasm exudes from the box, swells with water, and forms a new box of adult size normal to the species. This is called auxospore formation, and it is a peculiarity of the diatoms and some other plankton cells with the similar wall structure (figure 4B). The auxospore rejuvenates the diminishing cell and with its own peculiar surface structure forms anew the ornamented shell and, in many species, the long spines and other projections that the cell extends into the water.

Neither the auxosphore nor the adult cell is motile. It is known, however, that some of the centric diatoms, which predominate in the marine plankton and may assume almost miraculous complexity,

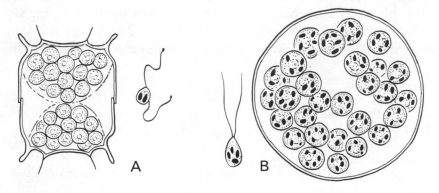

Figure 6. Zoospores forming in centric diatoms. A, *Biddulphia*, with the cell contents divided into two, each part forming thirty-two zoospores; × 500 (after Bergon). B, *Coscinodiscus*, with the cell forming sixty-four zoospores; × 400. (After Pavillard.)

form by multiple fission zoospores with two flagella. Since the adult cell is diploid, the zoospores may be gametes but the few researches have not yet proved the issue. From the purely botanical point of view there is still an enormous amount to be learnt of the centric diatoms, which are a culmination of marine phytoplankton.

In contrast the peridineans represent a culmination of motile plant-cells. They are often as large and elaborate as the diatoms, but they have flagella and the siliceous thickening of their cell-walls, though often as plates and spines, is not constructed on the principle of two overlapping half-boxes. One flagellum more or less encircles the cell equatorially and acts as a rotator or vibrator, and the second in animal fashion is directed backwards and propels. Both flagella

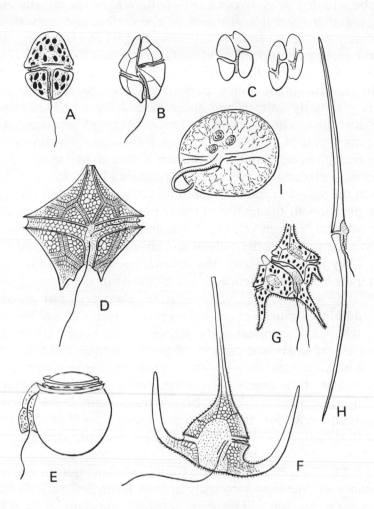

Figure 7. Marine peridineans (variously magnified) showing the transverse flagellum in a groove and the free longitudinal flagellum. A, *Gymnodynium*, a naked flagellate without cell-wall; ×500. B, *Gonyaulax*, with a wall of siliceous plates; ×500. C, *Gyrodynium*, dividing into two; ×500. D, *Peridinium*, with rigid armour of tesselate plates; ×500. E, *Phalacroma*, with the lower part of the cell much enlarged; ×500. F, *Ceratium tripos*, with the armour prolonged into three spines; ×500. G, *Ceratium*, showing the division of this complicated cell into two, each daughter-cell about to restore the original form; ×500. H, *Ceratium fusum*, elongated into two spines; ×250. I, *Noctiluca*, with finely striate tentacle, very small longitudinal flagellum, no transverse flagellum, no armour or cell-wall, no chromoplasts, but with three ingested food particles; ×100.

29

may be situated in grooves of the shell. Where the flagella emerge close together there is a soft spot on the surface and from this spot there may emerge in some species fine fingers of cytoplasm, which, like the stout pseudopodia of *Amoeba*, can grasp food particles such as bacteria and flagellates directed in vortices from the flagellar action towards them. Some peridineans feed therefore as plants, others are partly animal, and some indeed are wholly animal and have lost their chromoplasts. *Noctiluca*, the phosphorescent cell 0·5–2 mm. wide of northern seas, is such an animal peridinean, and it is big enough to eat minute crustaceans. Other animal species develop elaborate if elementary eyes and have stinging hairs, to capture prey, such as the jelly-fish and sea anenomes possess. Yet other species become parasites in the bodies of other peridineans, and some (called zooxanthellae) remain yellow–brown and apparently photosynthetic but live habitually in the cells of the animal plankton called Radiolaria. They seem not to harm the radiolarian, which may even benefit from their habitual presence. Some members of the animal plankton Foraminifera contain not yellow–brown plant-cells but green cells (zoochlorellae), which are related with the green *Chlamydomonas*. It is thought that the animal hosts of these plants benefit by absorbing the excess of sugar and oxygen of photosynthesis from the plant-cells, which, in turn, may use the carbon dioxide and nitrogeneous waste matter of the animal. The word symbiosis is used for this non-parasitic association of different organisms to imply living together in partnership; however, there is no satisfactory proof of this 'mutual benefit', though the partnership is well established. The peridineans therefore come fairly between botany and zoology. They are improved flagellates which are not completely enclosed in a cell-wall and can vary therefore into truly animal form, and this versatility, leading to parasitism and symbiosis, reveals the complexity of life in the surface of the sea: it has diverged into plant and animal and can re-form into a plant–animal. As living matter has specialized along these two main paths, the possibilities of recombination into symbiosis become less and less and those for parasitism intensify. Nevertheless there are at intervals striking examples from the combination of sponge and seaweed, and insect and fig, to the fields of farms and the pot-plants of town windows.

Little is known as yet about the life-histories of peridineans. The rather aberrant *Noctiluca* is the only form that has been found to have zoospores. The rest multiply by binary fission, which is generally nocturnal, but it is hard to believe that they have no other means of reproduction.

Both diatoms and peridineans may form rows of cells, held together by mucilage and resembling the filaments of cells that form the many-celled bodies of simple seaweeds. They lack, however, the coherence and polarity or direction of the true filament and are regarded not as real many-celled plant bodies but as colonies or loose

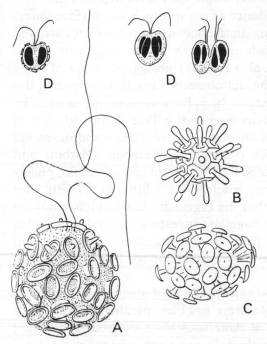

Figure 8. Coccolithophores, with yellow–brown chromoplasts. A, *Syracosphaera* (after Lecal-Schlander). B, *Rhabdosphaera*, skeleton. C, *Discosphaera*, skeleton. D, a simple coccolithophore, with two dividing cells; × 1000.

aggregates of single cells such as in the freshwater-biology textbooks are presented by the green plant *Volvox* and its allies.

The last group to be mentioned, because of their abundance in the sea, is that of the coccolithophores. Even less is known about them. For some time in fact they were known merely from the fragments of their skeletons dredged from the sea floor, and the living ones were not discovered until the voyage of HMS *Challenger* (1872). The simpler forms are motile and have two flagella, two chromoplasts, and a wall skeleton of discs or rounded lumps of calcium carbonate. The more elaborate are non-motile and have complicated skeletons.

They are mainly tropical, in contrast with the diatoms, which preponderate in colder waters, and the peridineans with an intermediate preference. They propagate by binary fission, but there is also evidence that zoospores can be formed, and some of the simple ones that have been described are undoubtedly the young flagellate, or larval states, of the non-motile.

The mineral shells, or skeletons, of these planktonic plants collect eventually, intact or broken-up in their passage through the digestive tracts of animals, on the sea floor, where they form deposits that may be characterized by the abundance of the species in the surface waters of that part of the ocean. Thus diatomaceous deep-sea oozes are well known, as well as others made up largely of the fragments of the coccolithophores. There are also on land geological strata of these organisms as, for instance, the diatomaceous earth (kieselguhr), deposited in seas of long ago whose beds have become dry land. The geological record of the diatoms goes to the Triassic period, that of the coccolithophores to the Cambrian, whereas the peridineans are known from strata no earlier than the Cretaceous. Diatoms and peridineans seem therefore to be late developments of oceanic plankton in contrast with the very ancient coccolithophores. But late development does not mean that the organisms have a higher state of evolution. It means that their flagellate ancestors proceeded to evolve into plant-cells at a later date than those of the coccolithophores, and no doubt, they vanquished other ancient groups of which we have no knowledge.

Already in the comparatively simple environment of the comparatively simple oceanic plant life there has been parallel and recurrent evolution. Different groups of flagellates have evolved at different times into similar but genetically different plant-cells as the termination of planktonic evolution. So, in the next stage, we shall find that radically different plant-cells have evolved into similar seaweeds through parallel and recurrent evolution in their environments, and radically different seaweeds have been modified in parallel recurrence into trees. We speak of a plant kingdom, as if all plants had one origin, but the kingdom now is a series of parallel lines of plants. If they diverged from a common ancestor, it must have been a very early pre-Cambrian flagellate.

Chapter 3

The Breaking Wave

SINKING INTO darkness a planktonic plant-cell dies, but sinking on to sunlit rocks where a shore rises, what may happen? This is how seaweeds begin. For reproduction they discharge zoospores, gametes, and non-motile plant-cells into the water. Coastal plankton consists therefore in its botanical part not only of truly planktonic flagellates and walled cells, which live their free lives, but of these reproductive units of seaweeds which have returned temporarily to a short planktonic existence before they descend and fix on to the rocks to become seaweeds. On the relatively narrow ledge of coast that lies between the spray-splashed upper limit of the shore and the beginning of perpetual night among the submarine rocks, there is the scene of botanical experiment into vegetation. No other part of the world has such diversity of plants. Green mingle with brown and red, and, as the microscope travels over their curious forms and investigation probes their silent lives, the great inventory of botany is revealed.

The reproductive cell, settled on the rock, begins to grow. When it divides by binary fission the cells stick together. Some microscopic seaweeds consist merely of a few cells, but in the longest, such as the oarweed, there are millions, adherent into a single being with a plant form. To reproduce they return the living contents of these cells into the water, while the plant body itself dies, loosens its hold, and drifts off to decompose. The many-celled plant body becomes, like the animal's, mortal, and its race survives by restoration to the plankton to start all over again. There are no seeds or bulky pieces for reproduction, but a true regression to the simplest state of their protoplasm. What wrought this change, whereby free-living cells came to attach themselves and grow into the threads, plumes, fronds, and tangles of the shore, was the ability to withstand and thrive in the increasing fury of the waves that they encountered.

If provided with enough light and suitable temperature, the

3

growth of the plant-cell depends on the supply of mineral salts and carbon dioxide in the water around it. An attached cell is at a disadvantage or at an advantage according as the water is still or moving. If still, the cell will deplete its immediate surroundings and the supply will depend on the slow and inadequate process of diffusion; if moving, the cell is bathed with fresh supplies. The greater the movement the better the supply and the better, or bulkier, is the seaweed, provided that it can withstand the motion, until, as on stormy headlands, waves may beat so fiercely that no seaweed can maintain itself. The relation between the size and structural development of seaweed and its position on the shore is a cardinal point in marine botany.

Waves approaching land alter their form. They travel the ocean as a swell, which is merely a gigantic ripple, with an up and down motion, gliding on; the cork bobs upward and forwards with the crest of the swell and downwards and backwards with the trough, returning almost to its starting-place when the wave has passed. These are waves of oscillation. As they approach the shore the shallowing water affects them. As they surge they gather water and at a certain depth, which varies greatly with the features of the shore but begins to be effective when the depth approaches the wavelength (measured from crest to crest), the bottom exerts a drag on water movement. The waves begin to rise, shorten, and quicken: there is a reining in. As the water shallows there is more delay in filling the crest and, when the depth of undisturbed water is comparable with the wave height, the surge upwards exceeds the wave speed. The crest begins to overtake the trough in front; it inclines forward and begins to break. The surge plunges and rolls up the shore, propelling the cork and even rocks landwards. The wave of oscillation has turned into a wave of translation by curling, combing, breaking, and tumbling forwards [9]. But the action has not ceased. After sweeping up the shore the spent water flows back and, as more is gathered for the next surge, it is sucked back in the undertow. Seaweeds therefore in deep water as at high tide heave upwards with the crests and downwards in the trough and exert a drag upon the swell. In shallow water they are pushed forwards and whipped back with its flowing. They live in the commotion where, except for the periods of exposure at ebb-tide, there is no possibility of stagnation. By their growths they display the varying rapidity of flow on the uneven coast.

The rocky shore is eminently suitable for plant growth, but the tide introduces a new effect by exposing the plants to air. It might be thought that this inclined them to become land plants but it is not so. Sunlight increases as the water ebbs, but the water supply is removed and the plants begin to dry. The higher the plant grows up the shore, the greater the quantity of light that it receives, but the less the water supply and the greater its desiccation. Down the shore the light-supply diminishes but the water supply increases and the desiccation is less. Below the level of low-tide, to a depth of 5–10 fathoms, the plant will still be within the reach of moving water, but light supply diminishes and begins to become a limiting factor for growth. At greater depths water movement will decrease and add another limiting factor until, at 50–100 fathoms, according to the clearness of the water, plant life will be unable to exist; in turbid water the limiting depth may be merely a few fathoms.

The shore is a much more complicated place for plant life than the open sea. There can be no wonder therefore that plant life established on the shore has become vastly more complex. The vegetation that we see at the present day as the tide goes out is the consequence of the rivalry of seaweeds that has gone on for hundreds of millions of years. They have become extremely involved, just as forest consists of many kinds of plant intermingled: tall and short, annuals, perennials, and epiphytes (plants growing, but not parasitic, on others). Nevertheless, the seaweeds sort themselves according to the environmental factors in a way that illustrates their evolution. It can be presented as a graph (figure 9), in which the effects of waves, tide, light supply, and exposure to desiccation are set against a section of the shore with its characteristic seaweeds. The size of the seaweed indicates its store of protoplasm and measures, of course, its success as a plant growth; and from sizes of seaweeds the suitability of different parts of the shore for plant growth can be judged.

The best place clearly is about the low-tide. The plants live here in the free run of the waves, where waves of translation are forming, breaking, or rolling according to the state of the tide, and where the undertow is strong. The very size of the plants carries them up to better illumination. Exposure is minimal and occurs merely for an hour or two at spring and equinoctial tides, and these are the rare occasions when botanists can get among them. It is the situation, on temperate coasts, of the oarweeds (laminarians), which have the most complex structure of all seaweeds and a complicated life-cycle (figure 24). On European coasts the larger reach 12–15 ft. long but in other parts of the world, which are worth a botanical journey for this reason only, they range into the largest of all plants. The giant kelp (*Macrocystis*) of the south temperate, rising from depths down to 80 ft., can reach the incredible length of 600 ft. Japanese oarweeds can grow to 90 ft. long in the space of a summer, at the rate of 6 in. a day [10]. The bull kelp (*Nereocystis*) of the Pacific coast of North America is said to produce in one year its stem 50 ft. long, topped by an air-bladder 6–12 ft. long, and surmounted by oarweed fronds as long again as the stem. On the southern coast of Chile occurs the sea palm

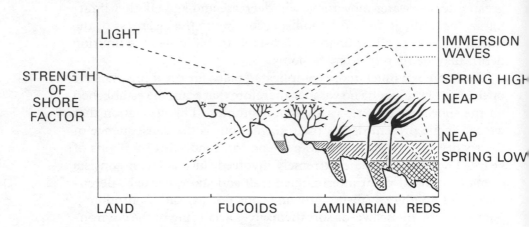

Figure 9. Graph relating the size and nature of seaweeds to the strength of the shore factors light, immersion, wave action, and tides. The largest seaweeds of strongest construction occur where exposure to air is least, wave action is greatest, and light is sufficient.

(*Lessonia*), which is a wonder of botany; its rigid stems, 6–8 ft. high and as many inches thick, stand erect with drooping fronds at low tide. Such are the conquests of seaweeds, measured by their immensity.

Higher up the shore in the region of neap tides live the bladder-wracks (fucoids) with lower stature, rather less complicated structure, and a simpler life-cycle. The shrubby *Fucus* may reach 4 ft. long, *Ascophyllum* 6–10 ft., and the sea thongs (*Himanthalia*) 15–20 ft. They are also massive plants whose bulk and firm texture add to the amenities of the shore for plant and animal life (figures 22, 23).

Above the zone of fucoids, about the half-tide level, exposure begins to tell. The seaweeds are smaller and usually of simpler construction. There are dwarfed fucoids, sea lettuce (*Ulva*, figure 13), its purple counterpart *Porphyra* (that slippery satiny membrane on the upper rocks), thread-like seaweeds, and encrusting patches. Yet higher in pools above the range of all but the highest tides there occur the fine fuzzy growths and minute threads of microscopic forms, attached to the rocks but living under the very special conditions of strong light, stagnating waters, and varying concentration as the pools evaporate and rains dilute. Little fingers, too, of seaweeds may hang on steep rocks where an occasional wave or uncertain spray may wet them. Up the shore therefore the seaweed life deteriorates and never becomes adapted to the land.

Below the laminarian zone in the sublittoral region, there predominate delicate brown and red seaweeds, membranous and filamentous, adapted to the dimming blue–green light. Some of them can be seen, too, in deep rock pools near low water, under the protecting fringe of laminarians that shelter them from the waves and shade them.

The lesson of the shore teaches that, in general, the greater the ebb and flow, the bigger will be the seaweeds. Shore zones are not, however, sharp, and some seaweeds, as *Ulva*, may occur throughout most of the littoral region. Nevertheless, laminarians do not grow near the high-tide limits; fucoids do not survive below the laminarians; and delicately constructed red seaweeds do not thrive in upper rock pools. In the long course of geological time, moreover, every kind of modern seaweed has ingrained into it a certain structure and manner of growth that now dictates its needs and its position on the shore. It seems as if their 'memories' were so long that they could acquire no more, and they do not alter their ways. It is not easy to state the requirements of any one because they concern not only the grown plant but its microscopic beginning. Clearly, however, sandy and muddy shores are proscribed. Shifting grains scour off the sporelings which start on them, though where wave action is slight, as in tropical lagoons, various green seaweeds (*Caulerpa*, *Halimeda*, and others) thrive on coral sand. On muddy shores, the fine shifting silt makes an unsuitable foundation and will smother the sporelings, yet a few red seaweeds grow in mud along the coast of Peru. Most seaweeds are confined to hard rock.

In temperate latitudes the spores of seaweeds are produced mostly during the late summer and autumn. They settle on the rocks, cleared

37

by autumn storms of the summer annuals, and they begin to grow at once. By March they may be an inch high. Then come the equinoctial tides with long exposure to the sun powerful enough to dry and to kill many of them; those that have started in unsuitable places will be so impeded that they are quickly overgrown. After their indiscriminate beginning, the March equinox, followed by the strong tides of a hotter April, is the decisive factor in parcelling out the shore to different seaweeds.

On land leaves are green. Other colours have no vegetative or growing significance. The young leaves of many trees, particularly in the tropics, may be white, yellow, brown, pink, red, purple, and even dark blue, and on opening from the bud they may hang in limp tassels reminiscent of seaweeds, but they soon turn stiff and green. The colours of flowers and fruit, which may vary much from species to species in a genus, are also not photosynthetic. Little importance is attached to them in classification; there are, for instance, blue, red, white, and yellow cornflowers (*Centaurea*). In the sea, leaves and stems may be green, blue–green, brown, or some shade of pink–red–purple, and these colours are highly significant because they are adjuncts of chlorophyll. They are chromoplast colours that work together with chlorophyll in photosynthesis. They can be likened to the sensitizers that are added to photographic films to make them react more quickly to weak light, or to react to light of a particular colour. The red pigment (phycoerythrin) of the red seaweeds absorbs especially light from the blue–green end of the spectrum, which is that which penetrates the sea most deeply and dimly. The brown pigment (fucoxanthin) of the brown seaweeds absorbs the yellow–orange light, complementary to chlorophyll, which absorbs particularly red and blue. The pigments enable the basic chlorophyll to operate photosynthesis effectively where, as in water, light is being weakened rather rapidly during its passage. All the pigments, moreover, are destroyed by the light at rates that depend on the strength and colour of the light and on the temperature and saltness of the water. Chlorophyll is the most resistant, phycoerythrin the least. Thus red seaweeds growing high up the shore in stronger light may be green, brownish green, or muddy purplish whereas the same species lower down is clearer red or purple, where the phycoerythrin is destroyed less rapidly and masks the chlorophyll. This susceptibility to strong light is one of the reasons why the equivalents of the brown and red seaweeds are not found on land: photosynthetically they could not compete there with the green plants, which, in turn, give way to the brown and the red in deeper sea water. Fresh water also destroys the brown and red chromoplasts and permits the colours to escape and bleach. Seaweed collectors, in making dried specimens by washing the saltness out in tap water, will observe how the red colour escapes most readily, and then the brown, so that oar-weeds may turn green and delicate red seaweeds with little chlorophyll become yellowish.

In contrast therefore with the colours of flowers, fruits, and young leaves on land, seaweed colours are fundamental. They entrain also a different chemistry. The sugars, starches, and oils, and the cellulose and mucilage-like substances compounded from the sugars, are different in the red, brown, and green seaweeds. Though all are made into similar threads, fronds, cushions, crusts, and shrubby growths, mortal and surviving by returning to microscopic spores, they present three ways of making plant forms, all based on the photosynthetic growth of attached cells but different in their chemistry. With these differences go also others in the microscopic structure, life-history, and flagellate character of the spores or gametes. Research has discovered that the three predominant colours indicate three separate lines of evolution among the seaweeds, and their common ancestor, if there was one, must have been a planktonic flagellate.

Both green and brown plant-cells and plant flagellates exist in the plankton, and both photosynthetic and flagellate characters of the spores of green and brown seaweeds relate them with such similar planktonic organisms. There must therefore be a brown lineage of plants from planktonic flagellate to oarweed, just as there must be the green lineage from such as *Chlamydomonas* to the green seaweeds and then to the vegetation of the land. To these must be added a much less successful yellow–green series, to which *Halosphaera* and some small or microscopic filamentous seaweeds belong. There is also a problematic blue–green series of planktonic and seaweed forms that have no flagellate form and which differ profoundly in protoplasmic structure, for they lack the vacuoles, chromoplasts, and highly organized nucleus of typical plant protoplasm. The case of the red seaweeds is peculiar. Their spores and gametes have no flagella, but their cells are typically organized, not as in the blue–green forms. They have no planktonic counterpart, for such as can be caught in the sea appear to be only their spores. By analogy it is argued that

39

they are the descendants of some ancient flagellate line, now extinct, which settled on the shore and specialized in its deeper waters.

Thus plant life of the shore is regarded as the outcome of these four kinds, at least, of photosynthetic flagellate lineages settling on the rocks and developing into as many plant forms. They are the green seaweeds (Chlorophyceae), the brown (Phaeophyceae), the red (Rhodophyceae), and the yellow–green (Xanthophyceae). To them must be added the exceptional blue–green seaweeds (Cyanophyceae). In contrast with the plant plankton of free cells, the seaweeds form the plant benthos of many-celled, attached, and mortal individuals. They begin the concept of vegetation [11].

The history of seaweeds is the story how these planktonic plants settled on the primeval shore and clothed it with plants, each line according to its photosynthetic method, yet all so similar in general effect that they have run almost exactly parallel courses. The temptation is to classify seaweeds by their shapes. It was the first way. For Linnaeus, in the eighteenth century, *Conferva* stood for the thread-like, *Ulva* for the frond-like, and *Fucus* for the more robust and shrubby. In the first half of the nineteenth century, there was a break-away and many genera were disentangled and sorted into colour groups. Then, as their life-histories became known, the colour groups were linked with flagellate characters and, at the turn of the century, the flagellate theory of seaweed classification, based on the construction and photosynthetic properties of their spores, was realized. Though seldom mentioned now, this reclassification of the seaweeds was a major revolution of botanical thought. It brought into the open an unavoidable conclusion. As plant life on the shore was the outcome of the attachment of plant plankton, the resulting plant forms and methods of reproduction, which have been carried on to land as the more familiar realm of botany, were not the product of one special line of inheritance, to wit one plant kingdom, but the consequence of any photosynthetic cell becoming attached on the shore and making good [12]. It demonstrated that the history of the higher many-celled plants was not so much to be deduced from the familiar plants of the land, which have inherited a great deal of their equipment from the sea, but it could actually be discovered in the great variety of seaweeds that enjoy the vigour of the breaking wave.

First efforts to classify plants are always artificial. Later, when more has been learnt about them, the classification is reshaped, and this reshaping goes on in continual refinement as the plants are better understood. It gradually becomes possible to assemble them in groups of increasing and fundamental likeness, not according to their superficial similarities. The ideal is classification by ancestral descent, but this can only be estimated from the true likeness of what plants are known; most have become extinct and are unknown. Nevertheless, on the principle that like begets like, the artificial are rearranged into what are called natural classifications. Unfortunately many of these are still largely artificial and the seaweeds are an example of the enormous amount of reclassification to be worked out in botany.

Chapter 4

Making Vegetation

THE SIMPLEST attached plant of the assemblage called plant benthos will be a single cell. Fixed to the rock within the range of light-penetration sufficient to maintain photosynthetic existence, water flows over it. Adherence is its new property. In the plankton stage fission and separation are the rule. Something, though it is not known what, must have gone wrong originally with the sinking cell that caused it to form an adhesive surface, and this failure secured it in the new environment. There are three ways in which it now fixes itself. The cell-wall fits the minute irregularities of the rock surface in microscopic dovetailing. The wall also sticks itself down with a glue formed by the condensation (or polymerization) of sugars as cellulose is formed, and which sets under water. The cell may also send out short tubes that fix themselves in the same way as extra anchors. Some microscopic seaweeds spend their lives in this state. Thus the temperate green *Halicystis* can become a sphere 1 in. wide, and the tropical green *Valonia* 2–3 in.; but, in doing so, these cells multiply the number of nuclei that they contain to very large numbers. Such sizes greatly exceed the planktonic and illustrate the advantages to plant growth that result from the attached, or benthic, existence.

Having reached full size, these cells extrude zoospores or gametes into the water and the cell body dies. How many kinds of plant there are of this sort, as a glorified cell, is not known because the microscopic are difficult to find. It is the way of attachment, however, in which all seaweeds start, and it is known that some microscopic forms that have been described as distinct species are in fact merely the beginnings of larger ones or a stage in their life-cycles.

The next problem is to consider how by cell-division into two, four, eight, sixteen, and more cells, they can be stuck together to make a plant body. In plankton, such division would lead to separation of the cells and multiplication of the individuals, but adherence to each other, after the adherence of the first cell to the rock, is the new principle on which they work and it multiplies the cells composing one individual. The cells are cemented together by adjacent walls, and their free surface to the water is covered by a film of mucilage. They cannot therefore move off elsewhere and, for reproduction, the protoplasm must escape from its housing as naked zoospores.

A group of cells may adhere rather loosely in a common sheath, or mould, of mucilage. The sheath is stuck down at the base and the cells lie in its substance. This is the habit of many diatoms, which live, not as plankton, but as fine fuzzy yellow and brown patches on

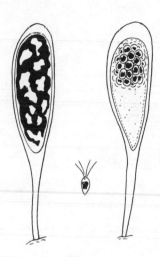

Figure 10. *Codiolum*, a microscopic single-celled green seaweed, living on rocks and other seaweeds, but possibly a stage in the life-cycle of the filamentous green seaweed *Cladophora*; the cell on the right forming zoospores; × 300. Zoospore; × 500.

rocks and indeed on the stalks and fronds of other seaweeds. There may be hundreds, even thousands, of cells in such a colony and, in a curious inexplicable way that the pennate diatoms have, the cells may glide about inside the mucilage. A structure so delicate is confined to quieter places of the shore, in rock pools and under banks of stout seaweeds, and it does not seem to have been the way in which the true plant body has built up. For this a much firmer union of the cells is necessary into a strong cord that can flex and strain in the waves, though mucilage still plays an important part.

Mucilage serves to strengthen the cell-walls, and it makes the slippery surface that lubricates the seaweeds as they glide over each other and over the rocks. As a soft layer, always washing away at the surface, it carries off the growths of diatoms and bacteria that would

darken the seaweed cells and impede their own growth. It is a condensation product of sugar-like substances, derived from photosynthesis, but it is not formed in fibrils like cellulose. It can absorb
water into its interstices and swell like a jelly until it disappears with
great dilution. It is present in all marine plant-cells and is another
property inherited by the seaweeds from the planktonic state and

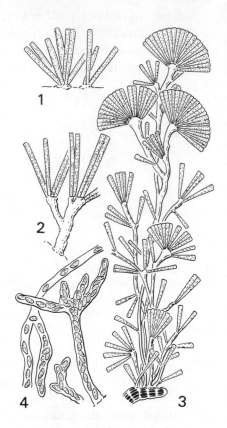

Figure 11. Attached pennate diatoms
forming tough mucilage strands on or
within which they raise colonies of cells.
1, 2, *Synedra*. 3, *Licmophora*. 4,
Encyonema. (From Oltmanns, 1922.)

amplified by them. As a carbohydrate it must not be confused with
the animal mucin or mucus, of similar consistency, which is a protein.

Plant-cells are strict in the direction in which they divide. The
partitions between daughter-cells are not laid down haphazardly.
They follow the direction in which the nucleus divides and the
partition wall is placed between the daughter-nuclei at right angles to
their line of separation. Generally, too, as shown by D'Arcy Thompson, the partitions occupy a minimum area between the divided cells

[13]. A cell may lengthen away from the rock or it may stretch side-ways over the rock; in both cases the walls between divided cells are transverse to the direction of cell growth (not necessarily to the direction of the rock). Some seaweeds have one manner of growth, others the other, while the massive fucoids and laminarians combine both. If the cell grows most actively on the free side away from the attached base – that is, if the cell grows apically – the new partition separates a basal attachment cell from an apical cell seated upon it

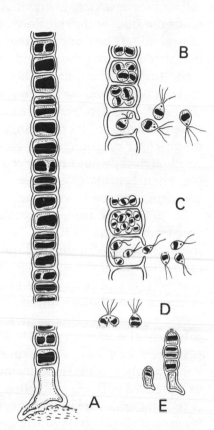

Figure 12. *Ulothrix*, a simple (unbranched) filamentous green seaweed. A, All the cells dividing except the basal hold-fast; the chloroplast (black) appears as a band or plate in each cell. B, Cells forming zoospores with four flagella and an eye-spot. C, Cells forming gametes with two flagella. D, Gametes uniting into a zygote. E, The growth of the attached zoospore into a filament; × 500.

and, if the process is repeated, a row of cells projecting into the water is formed; this is called a filament of cells. If, alternatively, the cell grows sideways in one direction and lengthens over the rock, a decumbent row of cells will be formed. If, then, these cells spread sideways other partitions will form to make a plate of cells. Similarly a plate of cells can arise through apical growth accompanied by

45

lateral growth from the cells of the filament, and a membranous frond, swaying in the water, will arise. Now a plant lying on the rock, though well illuminated, occupies much ground, and everywhere there is competition for footholds; the spores of seaweeds may settle in crowds on the rocks. But a plant projecting into the water has a minimum foothold, and in lengthening overshadows that which is prostrate, just as trees overshadow grass. The prostrate form is generally at a disadvantage unless it grows where projecting plants cannot; for instance, in crevices of rock pools where swirling stones scour the sides as the tide flows, or on boisterous headlands; but even these habitats are unencouraging for plant growth, the first being overshadowed and the second too rough. The projecting filament or frond makes the better use both of the available foothold and of the moving water, and by growing away from the rock it grows towards the light. This tendency is inherent in all vegetation, cramped from the abundance of its offspring. A seashore rock in spring is to the microscopic eye a forest of plants growing up into the water and so, one imagines, would have been the primeval rocks in the primeval sea when benthic plants had established themselves and begun to compete: the erect overcame the prostrate.

In projecting from its foothold, this simple filament or frond creates three new problems. It is the way now of attached plants that the more successful they are the bigger they become and, in occupying more space, the more are the problems that they have to meet and overcome. Plants come to command space because, as fixtures competing in photosynthesis, they must expose as much surface as they can for absorption, whether of sunlight or of substances dissolved in the water. Animals are compact and traverse space. The plant is a problem of maximum outside, the animal with its convoluted tubes of maximal inside; thus our sprawling factories with externalized guts are well called plants. The measure of plant success is the amount of space that it can occupy so as to oust its rivals. The biggest of all plants of which we have record must have been the old Indian fig of the ancients, under which Alexander the Great sheltered his army on their way into India.

The lead to this achievement from the microscopic compass of the planktonic cell has been the life of seaweeds. They pioneered plant form by sending from a fixed base filaments, branches, and fronds against the forces of the environment and so established vegetation. Progress was not direct but had to follow the system of generations

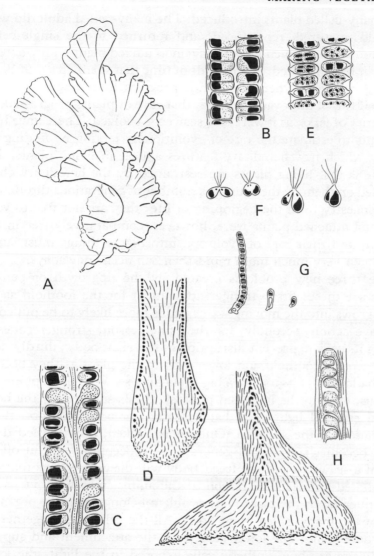

Figure 13. Sea lettuce *Ulva*, composed of cells like those of *Ulothrix* divided to form a large, membranous frond two cells thick. A, × ½. B, Frond in section; × 500. C, Cells near the base of the frond growing internally into hyphae descending towards the hold-fast; × 250. D, Hold-fast much thickened by the descending hyphae; × 80. E, Cells forming gametes; × 500. F, Gametes uniting in equal and in unequal pairs (two species of *Ulva*), each gamete with two flagella; × 1000. G, Zygote growing into a young *Ulva* by forming a *Ulothrix*-like filament; × 500. H, Base of the frond of the red seaweed *Porphyra*, composed of a single layer of cells, strengthened by hyphae formed on the outside of the cells; × 30 and × 80. (A–E, G–H from Thuret, 1878.)

47

that many-celled plants introduced. The many-celled adult did what it could, matured, reproduced, and returned to the single-celled planktonic state, which grew up again as a fixed plant with whatever ability it had inherited. At intervals of time that spanned a number of these generations, there was adult progress to a bigger or better plant; if this ability was inherited, that race of plants progressed. So in a series of jerks, as it were, the seaweeds evolved. There have been attempts to estimate the rate of evolutionary jerking by finding the rate at which new hereditary features appear in living plants. The trouble is that living plants are restrained by the hereditary check imposed on them by the countless number of generations throughout their ancestry from the beginning of life; they are not free to vary. The first attached plants were, however, comparatively free in this respect, and their rate of evolution into bigger things must surely have been very much more rapid than can occur nowadays.

The three new problems created by the upgrowth of benthic plants are these: first, stronger attachment for the foothold as the plant grows into the moving water and is more likely to be pulled off by wave action; secondly, for the same reason, stronger cohesion among its cells to prevent disruption of the plant body; thirdly, food supply to the foothold and lower part of the plant as they become overshadowed by its top. In a big branching seaweed these points are obvious, even to the impaired photosynthesis suffered by the hold-fast in reduced light. Yet that hold-fast, or adherent root, is the foundation of the seaweed. Once it is broken, the seaweed drifts away, becomes sterile, and generally soon decomposes. Cut off the root of a seaweed, and the freed body will die as surely as that of a decapitated animal. The reason in the animal may be obvious; no one understands the plant. The problem belongs with the organiza-tion of the many-celled plant body. All the cells work together for the whole plant. The lower cells form the attachment and support for the upper photosynthetic cells exposed to the light; the lower depend on the upper for their food supply, and the upper depend on the lower for their better station. The cells begin to specialize and lose their independence. The great principle of many-celled organ-isms, carried on into civilization, is the division of labour among the parts of the whole organism, or the making of artisans for the com-plex society of cells or men [14].

One way of strengthening the seaweed is to thicken the cell-walls. It is a rule in seaweed structure that the walls are thicker the nearer

1. A chloroplast of the tobacco-plant *Nicotiana* in section, shown by the electron-microscope; × 19,000.

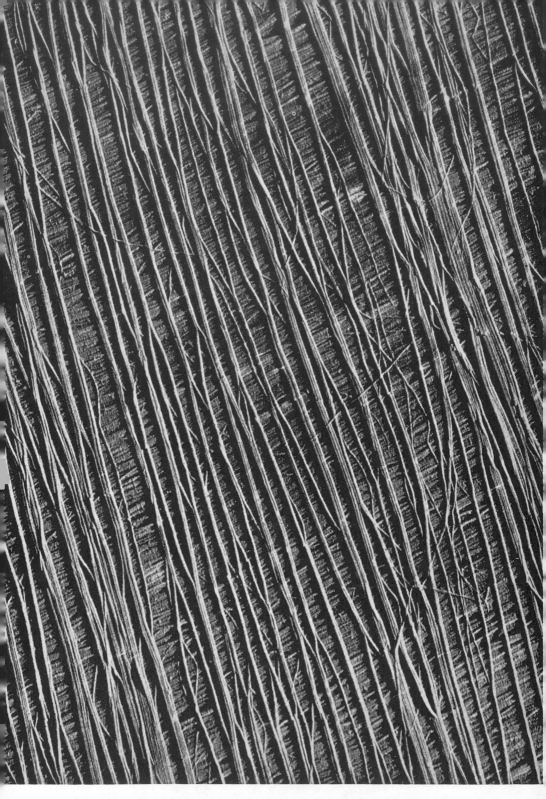

2. Cellulose laid down in two crossed layers of microfibrils in the fast-growing cell-wall of the *Ulothrix-like* alga *Chaetomorpha*; × 36,000.

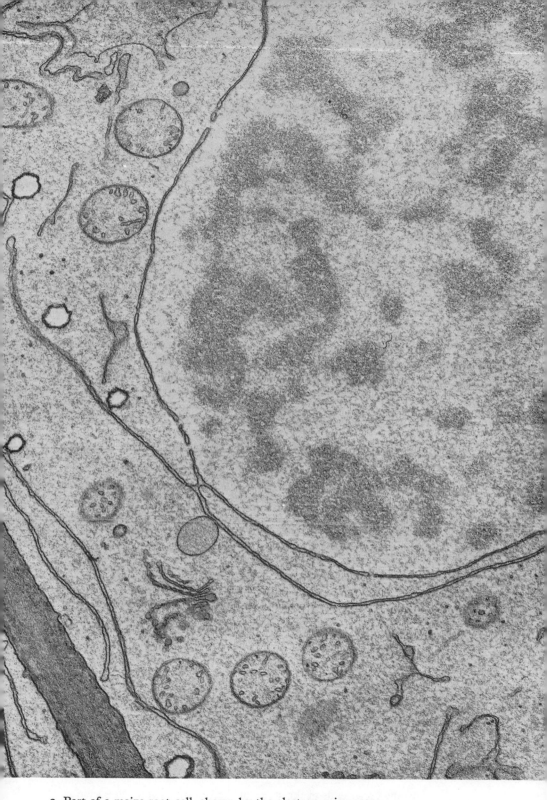

3. Part of a maize root-cell, shown by the electron-microscope.
The nucleus (upper right) surrounded by its fine double membrane
extending into the cytoplasm: the cell-wall (bottom left); × 17,000.

a

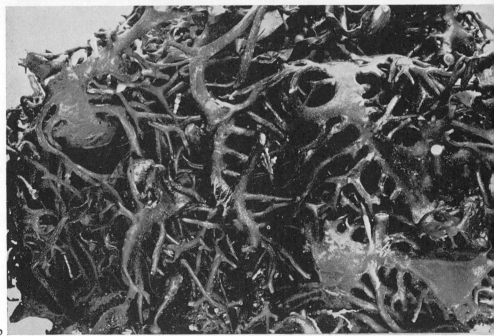

b

4. (a) the upper ends of the stems of the brown seaweed *Nereocystis* floated by the large air-bladders: Puget Sound, Washington; (b) The root-like hold-fast of the brown seaweed *Macrocystis*; California.

5. The brown *Lessonia* on the coast of Chile.

6. A dead tree bearing the brac[k]
of the polypore fungus *Fomes*,
in the swamp forest of Johore.

7. The lowland forest of
dipterocarp trees, about 200 ft.
high, in Trengganu.

8. The buttressed base of the Burseraceous tree *Santiria*, with stems of lianes; lowland forest in Johore with abundant undergrowth of small trees and palms.

9. Tree-ferns, *Cyathea contaminans*, in the mountain forest of Java.

10. The cycads *Encephalartos* in Transvaal.

11. Coniferous forest of Douglas fir *Pseudotsuga* and Sitka spruce *Picea* with little undergrowth; Olympic Peninsula, Washington.

12. (a) the nutmeg tree *Knema* with prop-roots on the compression
side of the trunk; swamp forest, Johore; (b) the Sapotaceous
tree, *Palaquium xanthochymum*, with stay-roots on the tension side
of the trunk; swamp forest, Johore; (c) the massive stilt-roots of the
screw-pine *Pandanus bidur*, a monocotyledon; Tioman Isl., Pahang.

a

b

13. (a) the lenticellate trunks of the swamp laurel, *Nothophoebe kingiana*, growing with the screw-pine *Pandanus*; riverside forest, Johore; (b) the buttressed base of the tree *Pentaspadon* (mango-family, Anacardiaceae), with roots undulated through variations in water-level; swamp forest, Johore.

14. The Tahitian chestnut *Inocarpus* (Leguminosae), with tension-buttresses along the main roots.

15. The lenticellate bark of the south-east Asian tree *Neesia* (Bombacaceae).

a

16. (a) the stout twigs of *Artocarpus anisophyllus* with pinnate leaves and axillary fruit-heads; (b) the stout twigs of *Artocarpus elasticus* with large entire leaves, and a pinnately lobed sapling leaf; see 28.

b

a

b

17. (a) the thinner twigs and smaller leaves of *Artocarpus lanceifolius* (right) and *A. fulvicortex* (left); (b) the slender twigs, small leaves, and immense cauliforous fruits of the jack-fruit tree, *Artocarpus heterophyllus*.

18. The papaya tree *Carica* of pachycaul habit with palmately divided leaves.

they are to the rock on which the seaweed grows (figure 15C); the thickness measures the strain set up by the to and fro motion of the breaking waves, as they surge and retire; there is no compression for a buoyant seaweed such as the land plant suffers under its own weight in air. Thick walls impede growth. They interfere with the absorption of mineral matter from the surrounding water and they restrict the swelling of the cell contents. Thus the plant body comes to enlarge not at its base, which would in any case weaken its cohesion, but at its tips where also light supply and water movement renewing the supply of mineral salts are maximal. The tendency to apical growth, away from the rock, so evident for a single cell in competition for foothold and for light, fits the mechanical needs of the larger seaweed by allowing the base to strengthen and the apex to remain thin-walled. So the familiar apical growth of plants finds its origin in the mechanical needs of the seaweed. It is inherited by land plants living under conditions so dissimilar that many of the most advanced, such as grasses and thistles, convert it as best they can into basal growth of the leaves.

Another way of strengthening is to lay down more walls by cell-division in more than one direction and to build thereby an internal and flexible scaffolding of intersecting partitions. This is the advanced method of parenchymatous or block construction found in all the successful brown seaweeds that stand the main force of the waves – the fucoids and the laminarians – and it, too, is inherited in all successful land plants as the way of body-building, adaptable to forces of compression. In block construction, not only do the cells divide longitudinally and transversely to make a mass of cells in three dimensions, but the partition walls are laid down in such a way across the cell that they knit together the contents of the two daughter-cells and their pre-existing walls. What is called a cell-plate forms between the two daughter-cells and extends outwards to the pre-existing walls. To judge from the size and the station on the shore of seaweeds built by this method, it is a stronger and better way than the filamentous, in which the partitions are formed by a ring-like ingrowth from the pre-existing walls to the centre of the cell. The block partition struts, but the filamentous hoops like a broad washer.

There are many and diverse details of nuclear and cell-division and of cell-wall formation in seaweeds that affect their construction, but they have been insufficiently studied and it is not known how the block method arose. All these variations are nevertheless the

4

experiments under the wave from which the best has been selected to dominate the shore and the land. The weaker survive in the inferior places where conditions are not so exacting but consequently less productive of plant growth: under the larger seaweeds, for instance, or as epiphytes upon them, in shaded and sheltered pools, and at the upper and lower stations of the shore.

In a row or filament of cells, each cell loses contact with the sea water on the sides where it joins its neighbours. Its absorbing surface is less therefore than in the single attached cell, just as this is less than in the freely suspended. In a plate of cells the absorbing surface

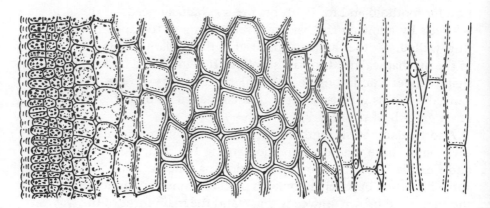

Figure 14. Outer part of a branch of the brown fucoid seaweed *Halidrys* in longitudinal section, to show the parenchymatous structure with small photosynthetic cells, actively growing and dividing at the surface, and the enlarging internal cells; × 125.

is further reduced by contacts with other cells all round. In a block of cells the absorbing surface is restricted to the outer face of the outer layer of cells, while the inner cells are both cut off from direct access to mineral salts in the sea water and shaded by the outer light-absorbing cells; the innermost cells are deprived of the external environment and come internally to occupy an inferior station. A weakening core does not seem suitable for extending seaweeds but the fact that they exist proves that the inherent disadvantages in parenchymatous or block construction have been overcome. This is the next step to be considered for on it depends the structure and success even of trees.

A block of cells always resolves itself in the seaweeds into three layers. There is an outer layer of small cells of planktonic size, $10-15\mu$ wide, deeply coloured with chromoplasts, intensely photosynthetic, actively growing and dividing: hence the maintenance of the small size which corresponds with that of planktonic flagellates as photosynthetic units. In the centre of the block there are comparatively few large cells, pulled out longitudinally, thick-walled, and practically colourless. Between the surface and the centre there is a gradation from the small photosynthetic cells to the elongate, non-photosynthetic, central cells.

If now the growing apex of a seaweed with this construction, such as *Fucus*, is examined it is found that all the cells are small and photosynthetic and that from this state the central cells are gradually transformed as the tissue grows older. The surface cells continue in active division, and thicken by this means the whole structure, while the inner cells are removed further and further from the outer sea water and light. They neither grow nor divide so rapidly, but are pulled out, and lose their photosynthetic pigments. For their maintenance they depend on their ability to absorb excess of food material as sugars and protein constituents from the photosynthetic cells. They exert therefore an inward diffusion, or movement, of organic food materials, wherewith they lengthen and thicken their walls. The lengthening accommodates the surface growth, which, as it were, racks them out, while the thickening of their walls strengthens the whole plant as a tensile cable, which, through the centre of the plant, takes the strain of the moving water. The surface cells divide in two ways, either by walls perpendicular to the surface, which increase the number of surface cells, or by walls parallel to the surface, which add to the number of interior cells. As more interior cells are thus formed so the inner ones are transformed gradually into the same form as the central, and the skeletal cable is widened. The parenchymatous plant therefore automatically differentiates itself into an outer photosynthetic cortex and an inner colourless skeleton, or medulla, and strengthens itself as it grows. The outer surface absorbs the light and inorganic food supply and the interior cells are supplied with elaborated organic food from the cortex.

This is the perfect organization attained by comparatively few, yet the most successful, seaweeds. It is presumably the result of selection of many attempts to improve the bodily, or somatic, growth. It is indeed complex and it depends on the differentiation of cells and the subordination of their individuality. Other seaweeds have some of

51

the necessary organization but lack other parts; they must form mechanically weaker, and therefore smaller, growths where the wave is less active. This is the reason why the connexion between seaweed structure and the place of the seaweed on the shore is such a fascinating study; it is the truly fundamental, lively and meaningful, anatomy of plants. Some seaweeds, for instance, cannot thicken their core as rapidly as surface growth requires. The central cells are pulled apart, and soft or tenacious mucilage fills the spaces to the detriment of their strength; they must occupy less favourable stations for growth. In other seaweeds the spaces fill with air and the weak body is floated: thus air-bladders, caused by the breakdown of internal construction, are made an asset. They should fill, presumably, with the gases of photosynthesis, particularly the less soluble oxygen of daytime, but the process is not so simple. The large air-bladders of *Nereocystis* contain as much as 12% of carbon monoxide. The peculiarities of these bladders afford a means of enquiry into the chemical work of the seaweed.

There is another method of strengthening the seaweed, which concerns chiefly the lower parts and the foothold. As the body grows this foothold (or hold-fast) must strengthen to keep pace with the increasing pull exerted by wave action. It thickens by outgrowth at the surface and so adds more internal tissue, as in the frond or stem, but as it is overshadowed this method is less effective. The hold-fast does not and obviously may not thicken by division of its internal cells because that would disturb its connexions with the rock. Instead, from internal cells, particularly from those where the inner cortex joins the medulla, narrow colourless branches grow out that force their way in the mucilage between the medullary cells downwards (figure 13C). They twine between the elongate cells, wrap them up in an irregular network, and fill all spaces, which tend to widen as the surface grows. They consolidate the central cable even further and they force their way into the internal tissue of the hold-fast, where they may convert its texture into an inextricable tangle of thick-walled threads; then some of them may burst through the surface and form new connexions with the rock. These threads are called hyphae after their resemblance to the hyphae, $2-10\mu$ wide, that compose the bodies of the fungi of the land. How the hyphae have originated their special form, function, and direction is uncertain, but they seem related to the down-growing branches, which, as will be explained, cover the main axis of filamentous seaweeds. They occur also in these plants of relatively simple construction, but they differ from the ordinary branchings of a filament of cells in that they are not directed upwards or outwards; nor are they photosynthetic but depend on organic food supplies within the plant, where they worm their way into regions of tension. They illustrate how the primarily photosynthetic and autotrophic plant, building its protoplasm from inorganic food supplies, becomes differentiated into an autotrophic skin and a heterotrophic interior supplied with organic food elaborated by the skin. The many-celled plant is bound to form an interior that gives up the primary peculiarity of the plant and becomes the skeleton.

This, too, is the well-known general construction of a land plant. One begins to see how it has come into existence, not by some imaginary strategy on land but by modification of the construction inherited from a marine ancestry, which acquired it in immediate relation with the environment. In land plants the long interior cells become the conducting tissues for water, which passes upwards in the central xylem, and for organic food substances, which pass downwards from leaves to roots in the peripheral phloem. Sea plants have no need of water conduction, but they have need of the downward conduction of organic food to the all-important hold-fast. What is obvious in giant seaweeds, where the hold-fast may work at several fathoms depth, applies also in small measure even to small filaments where the hold-fast needs more wall material, at least, than it can make itself. As the interior cells exert an inward pull on the organic food supply, the hold-fast must exert a downward pull and gradually, as the seaweed enlarged, this downward supply of food must have been improved. There is evidence that the elongate internal cells do conduct downwards but research into the problems of seaweed life, though so basic to the understanding of plants, is still in its infancy. It seems most probable nevertheless that, just as in the animal world the highest construction of sea life in the fish became the successful land vertebrate, the highest construction in seaweeds became adapted to the successful form of trees. Block construction, begotten and improved under the breaking wave, can be converted from a state of tension into one of compression by virtue of the thick walls, and its long cells can be adapted to water flow. Certainly none of the feebler filamentous constructions of seaweeds has succeeded in the green vegetation of the land.

Consider now the outward shape of the seaweed. From a single

53

initial cell a mass of cells is formed with the internal arrangement that has just been described. Swaying in the water or flexed rapidly from side to side and rolled in the run of the waves, this body will expand upwards and outwards as it shadows its base. If, for sake of argument, the body were to expand regularly in the geometrical form of an inverted cone, or even if it remained cylindrical, it would present a figure of least surface. Neither is a successful configuration when growth depends on surface absorption. Flattening into a frond-like form, expanding upwards, will give a maximum surface for the construction with strongly coherent cells. An open network with gaps between the cell arms, by which the cells adhere, will give the greatest surface, but the whole construction will be so much weaker as it lacks cohesion: yet even the network is exploited by filamentous green and red seaweeds in sheltered places. The frond-shape, spathulate, obovate, or reniform in outline, is the common and successful shape. It is what land botanists have called with contemptuous tendency, the thallus, to distinguish it from the elegant shape of land plants, tapering upwards in the conical outline of the shoot and downwards in that of the root. The thallus is, however, the shape of a photosynthetic organism growing attached and bathed in the water that supplies its food material: it is repeated with similar diversification in green, brown, and red seaweeds in their parallel evolution from single-celled beginnings. Once flattening has begun, however slightly through some inequality of surface growth, it will continue because the water will flow and eddy more rapidly at the edges of the thallus, where in consequence growth will be accentuated. Thus happen the frills, undulations, and even proliferations so commonly seen on the edges of seaweed fronds. The land plant enjoys no such congenial flow of food supply and, as will be considered later (Chapter 8), its thallus shape is modified. It is by no means coincidental that attached marine animals such as sponges, corals, and polyzoa resemble seaweeds in shape; the same principles of attachment, upward and outward growth, and exposure of a large surface to catch their food supply, apply to them. Plant shapes are inevitably evolved by attached organisms fed by the motion of the sea.

Simple seaweeds are filamentous. Some consist of a single row of cells up to one or a few inches long and all the cells except the basal hold-fast can divide and add to the length of the filament. Others are branching filaments that form the soft, green, brown, and red tresses in the water, a foot or two long. Being so much larger they need a stronger hold-fast, and the initial basal cell is thickened by down-growing filaments that clothe it in the manner of the hyphae of the parenchymatous plants. But the branching filament has established a manner of growth which persists through most forms of higher plant. It is apical growth. Instead of all the cells dividing by transverse walls and young cells with thinner walls making weak links in the chain, cell-division is restricted to the terminal or apical cell of the

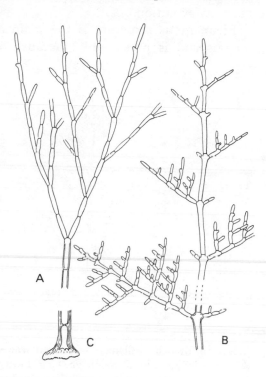

Figure 15. Filamentous growths of seaweeds, in green (*Cladophora*), in brown (*Ectocarpus*), and in red (*Callithamnion*). A, With similar and equal branches, but each branch with apical growth to give a succession of branchlets. B, Specialized with apical growth dominant into a main filament (with elongating cells) and short branches set in pairs, but developing only one branch of the pair in successive alternation; the branchlets developing on one side of the branches (secund habit of branching); the branch furthest removed from the main apex becoming a main filament in its turn. C, Hold-fast with much thickened walls; × 150.

filament: branching then arises by outgrowths from the other cells to establish lateral filaments with their own apical cells (figure 15). The whole structure grows apically at its tips, branches sub-apically, and progressively strengthens itself with thicker walls and hyphae down-wards. The apical cell somehow commands the daughter-cells that it has formed and prevents them from dividing, in spite of their growth, but it allows them at certain distances below it to branch outwards and form new filaments. Persistent growth by embryonic tips dis-tinguishes the plant body from that of the free animal, which after an embryonic beginning, followed in many cases by larval growth, completes itself; but the plant habit has been copied, or better evolved in parallel, by the corals, sponges, and other attached and plant-like animals of the sea. In land plants the connexion between apical growth and apical dominance is well known and is attributed to the effect of chemical growth substances that spread from the apex down the plant to inhibit branching. This control of growth, by which branches are spaced, has not been investigated in seaweeds, but they show its evolution.

Filamentous structure is the main alternative to the parenchymatous and is presumably, because of its simpler direction, the

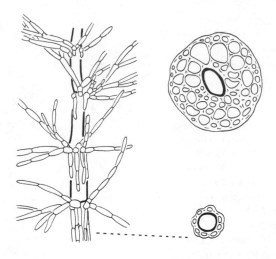

Figure 16. Branching filament of the red seaweed *Wrangelia*, showing the stout main axis invested by filaments growing down from the basal cells of the branches to build a cortex around the axis. Two cross-sections of the axis show the thickening of the cortex by the down-growing filaments, the outer remaining small and photosynthetic, the inner enlarging; × 40.
(After Feldmann-Mazoyer (1940).)

antecedent in the evolution of many-celled plants: indeed, nearly all the parenchymatous begin their lives, whether from spores or zygotes, by forming a filament, though it be microscopic. It is an unfamiliar construction in green land plants and in the sea it is nowhere so perfected as in the red seaweeds. There are simple unbranched filaments, branched filaments, corticated filaments, and

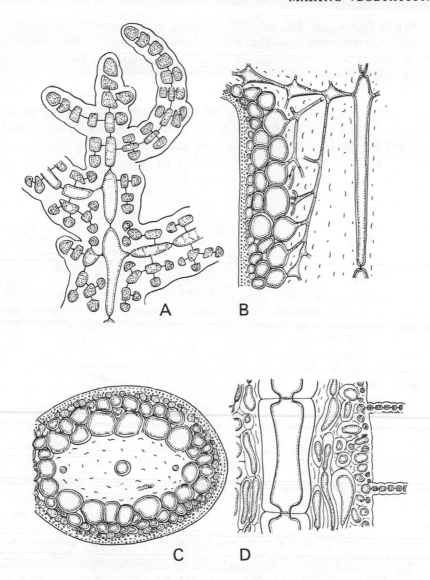

Figure 17. Cortication of filamentous seaweeds. A, Apex of the red seaweed *Bonnemaisonia*; × 250. B, Effect of cortication in *Bonnemaisonia* showing, in longitudinal section, a cell of the axis pulled out, surrounded by mucilage, and encased by the cortical filaments with minute photosynthetic surface cells; × 50. C, Same in cross-section. D, Compact cortication of the axis of the red seaweed *Plumaria*, the inner cells of the cortex having developed hyphae; note, also, the surface hair like the filament of *Ulothrix*; × 50. Observe, in A, as in figure 18C, the very precise growth of the filament.

57

multiple filaments, the last two showing in a characteristic way of plant improvement how the whole system of branching filaments consolidates upon itself and produces a result similar to the parenchymatous.

With the corticated filament, instead of the branches lengthening outwards and repeating the form of the main filament, they remain short and more or less cover the main filament with a cortex of short photosynthetic filaments. The cells of the main filament then generally elongate and become thick-walled to form the skeletal axis,

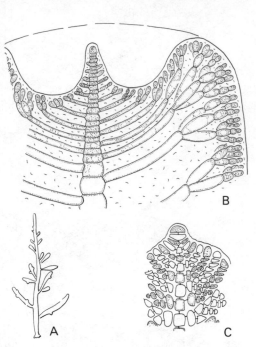

Figure 18. Cortication of the filament with all the branches enclosed in a sheath of mucilage. A, Clumsy plant of the red seaweed *Chondria*; × ½ (after Oltmanns, 1922). B, The elegant apex of *Chondria* in longitudinal section; × 70 (after Falkenberg, 1848). C, Apex of the frond of the red seaweed *Membranoptera*, showing how (in alternate stippling) the filaments, restricted to one plane, are built into a frond; × 300. (After Kylin, 1922.)

which is often invested with the downgrowing hyphae from the inner cortical cells. Furthermore the whole filamentous structure, condensed on the axis, becomes invested by a continuous coating of mucilage, which, if tough, gives the structure the appearance of being embedded in a strengthening yet transparent plastic. Some of these seaweeds develop equally all round and form slender cylindrical or branched growths like stout simple filaments. Others have the branching restricted to one plane, from opposite sides of the main filament, and, when the branching is compact and encased in the common mucilage sheath, they form flat fronds of one-cell thickness

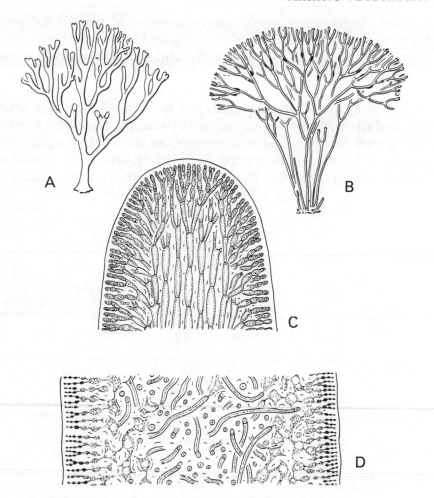

Figure 19. Seaweeds composed of many branching filaments encased in a common mould of mucilage (the multifilamentous structure). A, Green *Codium* or red *Nemalion* or *Scinaia*, and B, the red *Polyides* (with spore patches, in black); × ½. C, Apex of a branch in longitudinal section; × 300. D, Cross-section of such a mature structure in the thick frond of the red *Dilsea*, with small photosynthetic cells (in black) and hyphae developed in the internal mucilage; × 75.

except where the main filament is thickened by hyphae into a midrib. Such are the beautiful red fronds of *Nitophyllum* and *Delesseria*, by no means so delicate in the water as they seem when made into herbarium specimens. The fronds of *Delesseria antarctica* are 1–2 ft. long. Among the brown seaweeds *Desmarestia* has a highly branched feathery thallus up to 10 ft. long, consisting of narrow flattened branches built on the system of the corticated and encased main filament; it grows in the colder northern and southern seas below the laminarian zone.

In the most complicated multifilamentous structure there are several to many longitudinal filaments in place of the single main one, and all their branches are encased in a common mucilage sheath. This sheath may be tough, as in the rubbery worm-like purplish-red fingers of *Nemalion*, which hang from the vertical rocks of stormy headlands where few other seaweeds survive. Then if in this mucilage

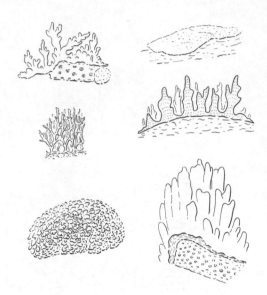

Figure 20. Various kinds of nullipore, or calcareous red seaweed, pink to carmine-red, growing in their stony and obtuse way on dead coral; $\times \frac{1}{2}$.

sheath calcium carbonate is deposited as an additional skeleton of lime, similar to that of corals, the tough calcareous seaweeds are produced, which are so characteristic in green and red forms of tropical seas. It is said indeed that the calcareous red seaweeds, which form pink and red stony fingers, antlers, knobs, and cement-like encrustations, are often if not always the main agents in consolidating coral reefs, both on their landward and seaward fronts [15]. Deposits of calcium carbonate, even microscopic, are fairly opaque and as these seaweeds thicken their inner parts become heavily shaded; of all seaweeds, their growth is peculiarly superficial. Yet because of their sensitive red pigment they can grow in the clear

water that corals require even at depths of 60 fathoms. On the landward side of the reef they grow over broken pieces of coral and cement them into platforms and curving ledges. On the seaward side,
where surf beats incessantly to make the most energetic of plant
habitats, they intrude between the dead bases of the corals with
flanges and fingers, and their own dead and stonier parts unite the
coral blocks into continuous rock. Corals have small pores over their
chalky skeletons through which the animal polyps protrude. Calcareous seaweeds, often resembling the finger-like corals, have no
such pores and they are often called in consequence nullipores. In
temperate waters they are represented mainly by slender branching
corallines (*Corallina*) and the cement-like *Melobesia*, which forms
bright pink encrustations over the bottoms of rock pools shaded by a
fringe of *Fucus* or *Laminaria*.

A final method of strengthening the corticated filamentous structure, whether multifilamentous or not, is by the strong swelling of
the internal cells, which press tightly against each other and give
with their thick walls a turgid and more rigid effect (figure 17B, C,
the cortex). At first sight under the microscope this structure seems
to be parenchymatous and it is given the name of false parenchyma
(pseudoparenchyma). Its real nature can be made out in two ways.
First, it is built up by filamentous growth, which can be seen at the
growing apex. Secondly, being filamentous or composed of rows of
cells, there are minute pores in the transverse walls between adjacent
cells of a row but not between adjacent cells of different rows. By
studying the occurrence of these pores with the strands of cytoplasm
passing through them, the manner of branching of the filaments
composing the structure can generally be made out without difficulty. False parenchyma is typical of the larger and tougher red
seaweeds such as *Rhodymenia* and the more tropical *Gracilaria*, both
of which can grow in fairly boisterous places. It reveals in imitation
the superiority of true parenchyma. False block structure is the
culmination of filamentous seaweed structure. It does not occur
among land plants except in the fructifications of fungi, such as the
cups, morels, toadstools, and puff-balls of vegetable decay, about as
far removed from the breaking wave as any plant habitat could be.

Chapter 5

Living Seaweeds

THE OVERFLOW of plankton brought plants down to bed-rock. That meant a new life or extinction. To succeed, the plant-cell had to conform with the new conditions, which meant attachment, building a many-celled plant body with a certain microscopic structure, shaping it into plant form for absorption, and strengthening it. No matter what its origin in the plankton, the cell on the rocks had to follow these rules. Today this is revealed to us by the three great series of green, brown, and red seaweeds: in parallel, independently, they have produced similar results, though some have achieved in one way or another more than the rest. The making of the larger benthic growths can be followed in the development of the large seaweeds. The limitations of the smaller and less successful can be understood from their stations on the shore.

How this happened two or three thousand million years ago – and a hundred million is neither here nor there with such numbers – on the primeval coast of the pre-Cambrian period, by what plants, in what space of time, or in what place, may never be known, but the ruling conditions for plant life cannot have been much different from the present. When the tide recedes tomorrow there will be exposed again a scene of enormous antiquity where plants work according to their talent in situations as humble as the most primitive, as diverse as there ever were means of progress, and as high as marine plant life can attach, because vegetation is by now fulfilled. There will be exposed the tableau of a great epoch of plant evolution until – and man must decide – it is overcome with the sludges of civilization. Living seaweeds are the modern actors of the old drama. At first it was simple but it may have attained modern complexity a thousand million years ago, since when the cast has been changing, for there is evidence from the distribution of some living seaweeds that they cannot have existed so very long ago. The salt tang is a primitive tingle; the ebb-tide must beckon the botanist. The first peep into a rock pool at low water is so inspiringly beautiful that the vision never fades. Yet when they explore, as explore they should, for that is the only way of learning, botanists penetrate inland and forget the sea, which holds the answers to so many riddles of their science [16].

In clothing the rocks the seaweeds have altered the shore. Whether floated by the rising tide or prostrated in its falling, they shade and shelter. Under, around, and upon them spring up new situations to be exploited by other seaweeds and by animals, and they in their turn create their situations for smaller organisms. This is the effect of sedentary, attached plants. They build, and into their colonnades animals pursue; in their recesses, animals lurk. Every growth makes as a fixture for some time, a micro-habitat with a micro-climate where micro-organisms survive. Survive? That is the complexity of the modern shore. Under and upon the dominating plants of ultimate achievement there are all sorts of survivors from times when plant evolution was incomplete. Shore studies must reckon with them and sort them out according to the special conditions where they survive; where, for instance, microscopic plants prevail, where filaments prevail, and where there are simple fronds or cushions, as well as giant oarweeds. If a section of the shore, as in figure 9, introduces the main progression of plant form and structure in relation to the environment, another and a bird's-eye view is needed at right angles to discover the secondary effects [17].

The plan in figure 21 shows a bay with headlands, a stream, its silt, the sand, the rocks, and the disposition of the marine plants. Waves entering the bay are deflected by the shelving floor towards the head-lands and spread around its recesses. They break strongly on the headlands with turbulent eddies, less strongly on its flanks where the reflected waves chop into the commotion, and they are dispersed towards the head of the bay. Sand collects where the wave action is weakened, mud where it is least. Here seaweeds rarely obtain a foot-hold, but if rocks project where the stream scours, certain thin kinds (e.g. *Fucus ceranoides*) adapted to less saline water, will grow on them. In contrast, only the toughest will grow on the tidal rocks of the headland. If the headlands plunge, there will be a steep narrow sublittoral zone for the growth of strong seaweeds. The widest stretch of clean shelving rock lies on the flanks of the headland, for up-shore eddying waters will deposit silt and sand; the most luxuri-ant growth of seaweeds is on the seaward part of the flanks. Rocks near the head of the bay will have a fine felt of short seaweeds, the filaments or slender strands of which can grow around the sand grains that settle on them. Sea birds will favour certain ledges, and the rich rain water that has flowed over their droppings will promote the growth of some seaweeds, such as the sea lettuce (*Ulva*), and impede that of others. Everywhere will have its special situations. Every bay will differ according to its rocks and the way they decompose into shelves, clefts, ridges, and hollows to form rock pools

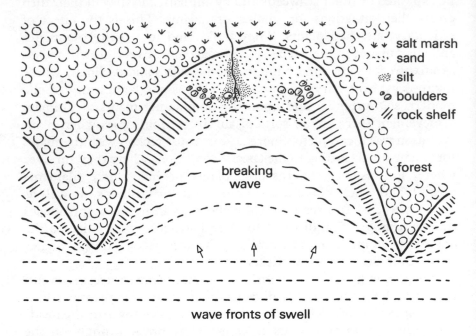

salt marsh
sand
silt
boulders
rock shelf
forest
breaking wave
wave fronts of swell

Figure 21. Diagram of two headlands and enclosed bay, to show the spreading and breaking of the wave, and its deposition of silt, sand, and boulders.

at different levels, all with their particular conditions of light, water supply, and exposure. All these differences will be reflected in the plant and animal life, for no organism is so accomplished that it can live in the face of competition with others everywhere: every plant and animal belong to a species, that is a certain manner of living.

Then if the coast is one where currents set, waves will approach at an angle. One side of the bay will be more sheltered than the other, and the vegetation of the one will differ in species or in abundance from the other, and this will be reflected in the animal life which will react upon the plant life. According to latitude there will be seasonal differences and temperature differences. In high latitudes, ice grinds young seaweeds off the rocks. In low latitudes exposure to intense sunlight and desiccation kills those that may start on the upper part of the littoral zone. A cloudy coast may be better than a sunny, yet a monsoon season will kill delicate seaweeds with too much rain. Lastly, geological events in the past have separated land masses and united others: intervening land and sea will act as barriers to coastal dispersal: the northern and south sides of a continent, being in very different latitudes, will separate the east and west flanks, and subsequent geological events will variously unite these distant flanks with other land masses. Different seaweeds will evolve in the long-severed parts of the earth. There will be therefore, besides the environmental distribution of seaweeds on any one coast, a geographical distribution according to latitude and climate, and a geological distribution according to their history. The fixed world of plants is immensely more complex than the drifting world of plankton [18].

Temperate coasts are much richer in seaweeds than tropical. Among the more notable are those of the Pacific border of North America, of Chile, South Australia, Tasmania, New Zealand, Japan, France, and indeed the British Isles, where Cornwall and County Kerry are the richest. The tideless Mediterranean and the shallow Baltic with its large influx of cold fresh water from rivers and melting ice are among the poorer. In the tropics, where the temperatures of sea water range from $18°$ to $28°C$, seaweeds have to compete with corals and, where the sea is clear and running freely, corals generally win. They cover the rocks of the lower littoral and sublittoral parts of the shore with strong growths on which seaweed spores cannot sprout or, if they should, cannot survive. Living corals seem to be strongly antibiotic and inhibit the growth of other organisms upon them. Yet, some of them have in their tissues vast numbers of symbiotic plant-cells like the zooxanthellae of the radiolarian. They have become dual organisms, animal by nature but plant by intake of symbiotic cells and marine effect. Nevertheless, as in all cases of symbiosis, the exact chemical interrelation between the partners is neither clear nor convincing of the philosophical notion of symbiosis. The zooxanthellae do seem, however, to remove the nitrogenous and phosphatic excretions, which are the surfeit of animal feeding [19].

On the upper part of tropical shores intense sunlight and evaporation, caused by the heating of the rocks, are as destructive to corals as

65

to seaweeds. The corals thrive in the first tumult of the waves. With the help of the nullipores they advance their strong shelf seawards to build up the reefs that fringe most rocky tropical coasts. Behind the belt of vigorous growth the corals begin to die. Such as are not cemented together by the nullipores are slowly pounded and triturated into the coarse sand of the lagoon. Occasional violent storms

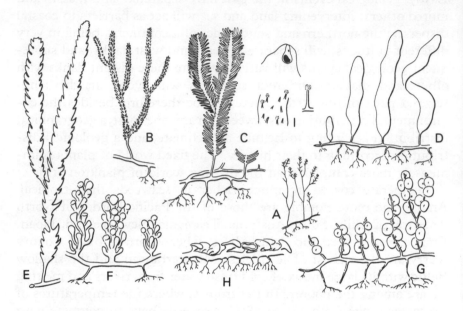

Figure 22. Species of the green seaweeds *Caulerpa* of warm temperate and tropical seashores; × ¼. These seaweeds are composed of tubes, 0·1–3 mm. wide, lined with protoplasm containing many thousands of nuclei, but not divided into cells. A, *C. fastigiata*. B, *C. cupressoides*. C, *C. sertularioides*. D, *C. prolifera*. E, *C. scalpelliformis*. F, *C. racemosa*. G, *C. macrodisca*. H, *C. nummularia*. I, Spikes from the frond of *C. prolifera* through which the zoospores escape; zoospores, × 500.

and tidal waves from earthquakes hurl large blocks of coral into the lagoon, where they die and then become footholds for seaweeds. Many delicate red and green seaweeds live on the sheltered undersides of the dead coral lumps in the light reflected from the brilliant sand. Many peculiar green seaweeds grow also in this coral sand where it is little disturbed by waves in the lagoon. Here are found the green sea grapes (*Caulerpa*). Their stems creep over the sand and root by fine branching threads. They bear, according to their species, a great variety of fronds, fern-like, moss-like, grape-like, and fila-mentous. The internal structure is peculiar because it contains thousands of nuclei scattered in the peripheral cytoplasm and an irregular internal skeleton of cellulose struts which have no connexion with nuclear division and are not cell-walls. *Caulerpa* is, in fact, an elaboration of the many-nucleate construction that starts with such simple globose forms as *Halicystis* and *Valonia*; and this series of green seaweeds shows how plant forms can be evolved without nor-mal cell construction in parallel to the normal. Thus, allied with *Caulerpa* and found in the same places, is the jointed green calcareous alga *Halimeda*, built on the same principle of many-nucleate tubes, but now of multifilamentous construction in parallel with the jointed red corallines.

In these coral lagoons there may be found, at least in South-East Asia, one of the most curious organisms. It forms a tuft of stiff, olive-green, branching fingers up to a foot long, on which appear shallow holes as though it were a green sponge. In fact it is a red seaweed, permanently discoloured, and consists of many fine intertwined branches of multifilamentous structure forming the fingers, over which grows a thin film of a true sponge. It is another dual organism but this time of green red seaweed and sponge. Similar growths have been called *Thamnoclonium* and *Ceratodictyon*, but I have not suc-ceeded in naming this one and have never met another person who seemed aware of it. How many more such demonstrative oddities exist on the little-explored shores?

In colder waters, the green seaweeds form a slippery zone near high-water mark where the filmy fronds of the sea lettuce (*Ulva*) grow with the slender strands or flattened tubes of *Enteromorpha*, so common on sea walls. Elsewhere, on temperate shores, green sea-weeds are scattered throughout the tide range in the form of tufts of branching filaments (*Cladophora*) and as the dark olive-green spongy cushions and dangling fingers of *Codium*, in structure similar to *Halimeda* but not calcified. In fact nowhere do green seaweeds dominate the shore in the manner of the brown and the red [20].

Red seaweeds form the underworld of shore plants, brown sea-weeds the upper world. The brown dominate all coasts except the tropical, and they excel because of the parenchymatous construction of the fucoids and laminarians. Though these two are undoubtedly related because of their photosynthetic pigments, metabolic pro-ducts, and flagellate state, they nevertheless differ greatly, and the common ancestor to them seems to have belonged to the very distant past. The fucoids grow apically and can assume forms with stems, branches, and leaves like brown editions of land plants. The laminarians grow at the base of their broad fronds where they join the stem, and their hold-fast, unlike the usual disc-shaped attachment of the fucoids and other seaweeds, has been improved into a root-like system of forking branches strongly adherent to the rock, and able to increase the attachment to a foot or more in diameter as the fronds enlarge or multiply and the pull on the plant gathers.

The bladder wrack (*Fucus*) of northern coasts is a flattened, forking seaweed, the apex of which divides into two equal parts; that is to say, it dichotomizes. Each half becomes a branch but generally the branches grow unequally. One grows faster and overtops the other before dichotomizing again. The shorter branches are set alternately, with the effect that the alternate longer branches build up a slightly zig-zag main stem on which the shorter appear as if they had been later outgrowths. This process of alternate overtopping to build up a median stem with side branches is widespread in seaweeds and is inherited in land vegetation, where it can be detected in the pinnate leaves of ferns. But *Fucus* is a simple member of its group, its only elaborations being a midrib thickened with hyphae and, in some species, air-bladders in the thin wings of the branches. The air-bladders contain oxygen (15–30%), carbon dioxide (0·5–2·5%, the quantities of which depend on photosynthesis), and a remainder of nitrogen.

More advanced fucoids elaborate the method of overtopping and specialize the short branches thus formed. The growing point is situated in an apical notch. The halves of the apex destined to be short branches, formed alternately from the two sides of the main apex, remain dormant in their small notches for a considerable time, for they are suppressed apparently by the influence of the main apex. During this time the subdued apices can divide and initiate several short branches that grow out eventually into short leaf-like structures, stalked air-floats, or fertile branches. Then the main apex can itself undergo equal dichotomy and produce two equal main stems that continue with unequal dichotomy and overtopping. Thus are built up the complicated plants of *Ascophyllum* and *Halidrys* on British shores. In contrast with them are the sea thongs (*Himanthalia*), which are simple dichotomous thongs developed from a button-like base that is the first year's growth of the plant.

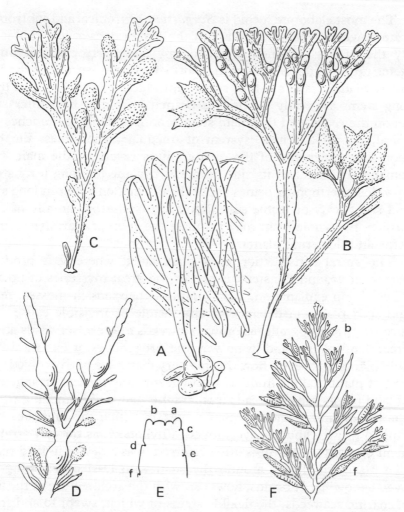

Figure 23. Brown fucoid seaweeds. A, Sea thongs *Himanthalia*, with the button stage (first year) and central outgrowth of dichotomous straps (autumn of first year, and the second year); × ⅙. (After Newton, 1931.) B, *Fucus vesiculosus*, the bladder wrack, equally dichotomous, with fertile tips, and air-bladders; × ¼. C, *F. spiralis*, to show alternate overtopping of one branch of a dichotomy to form a main axis; × ⅙. (After Thuret, 1878.) D, *Ascophyllum*, showing equal dichotomy of the main axis and the development of short sterile and fertile branches; × ½. E, Diagram of the apex of a main branch of *Ascophyllum* to show the dominance of the central apex and its successive divisions to form branches (a–f), which by overtopping become the short branches. (After Oltmanns, 1922.) F, Sea oak *Halidrys*, as a more elaborate *Ascophyllum* with short sterile branches, air-bladders (b), branches, and fertile branches (f); × ½. (After Oltmanns, 1922.)

The most elaborate fucoid is *Sargassum*, of tropical and subtropical waters, where it takes the place of the laminarians at low-tide level. By the process of unequal dichotomy and overtopping it builds stems up to 12 ft. long, set with short shoots consisting of a flat leaf, an air-bladder, and a tuft of small reproductive branches. But these long stems are really themselves branches from a short thick stem set on the rocks, and this main stem produces the long branches in a spiral, not an alternate, system of unequal dichotomies. Furthermore, the inequality of the dichotomy is reversed; the main stem remains dwarfed and its side branches become the long leafy stems in which overtopping brings about the distinction between long stem and short leafy or fertile side shoots. In *Cystoseira*, an ally of *Sargassum*, the whole plant may simulate the stem–leaf arrangement of ferns and flowering plants.

The spiral construction of a plant-apex, whereby it produces leaves set around the stem, is one of the great mysteries of botany. There is no explanation. Very simply it happens in mosses, ferns, and seed plants in defiance of any visible or invisible cause. The fucoids, though parallel as brown seaweeds to green seaweeds and to green land plants, and in no way the progenitors of either, are the only plants that reveal how this mystery may have been evolved: it is part of plant-making in the sea, and in this way the leaf arrangement of land plants must be read; it is an inheritance of branching from a spirally organized dichotomizing apex in which the branches have limited growth and are overtopped. Phyllotaxy, as the leaf arrangement of plants is called, is often referred to as a device in land plants that places the leaves around the stem with least overlapping and overshadowing. It begins, however, with the architecture of the apex of marine seaweeds, the flexible stems of which, swept to and fro in the water, present no such problem of overlapping. If ever proof were needed of Church's dictum that 'the beginnings of botany are in the sea', it is in this occult problem [21].

Sargassum, which should be famous on account of its structure, is better known as the cause of the Sargasso Sea which so troubled, and still may trouble, the sailing ships of the north tropical Atlantic. From longitudes 27° to 80° W and from latitudes 18° to 35° N, between the Azores, the West Indies, and the south Atlantic coast of the United States of America, there float in vast numbers over many thousands of square miles bits and pieces of the brown leafy stems of *Sargassum*. In places the plants are scattered, in others so densely entangled, just breaking the surface, as to clog the passage of the sailor. It is the only seaweed that habitually thrives adrift. The air-bladders (a quarter to half an inch wide) keep it buoyant, and in warm waters, never less than 18°C, the weeds continue to grow vegetatively. About a dozen forms can be recognized from the size and shape of the leafy stems, but they intergrade, and some indeed are recognizable as bits of West Indian species of *Sargassum* that grow on the shore. It seems that the Sargasso Sea is composed of

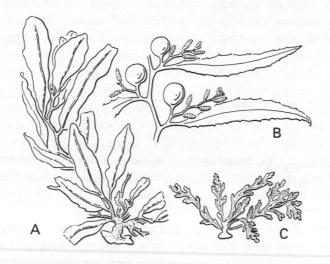

Figure 24. Brown gulfweed *Sargassum*. A, Base of a plant to show the short, stout main stem with the lengthening and leafy branches set around it; × ½. (After Setchell, 1933.) B, Part of the leafy branch of another species, with axillary air-bladder and fertile branches; × 1. C, Young plant of *Cystoseira* to show the short main stem with the branches borne spirally round it in phyllotaxis and transformed by overtopping into fern-like leaves; × 1. (After Oltmanns, 1922.)

Sargassum plants of the coasts of the Guianas, West Indies, and, perhaps, West Africa, which have been torn off by storms or nibbled off by shore animals, to be drifted out to sea and caught in the Gulf Stream and the counter-equatorial current from West Africa (figure 24). Probably this multitude of plants, not primarily but secondarily planktonic, forming the biggest stretches of uniform vegetation in the world is the accumulation of geological time. Accompanying the weed are shrimps, crabs, molluscs, and small fish, which thrive on the weed and each other as well as on hydroids and polyzoa that grow on the weed. All of these reproduce sexually but the *Sargassum*, like all detached seaweeds, remains sterile. According to high authority the very ancient and polyzoan-like organisms called the graptolites, found in Palaeozoic rocks, lived attached to floating seaweeds; they may indicate the beginning of Sargasso Seas [22].

To survive detached, though sterile, is peculiar to several fucoids. *Fucus*, *Ascophyllum*, and *Pelvetia* can survive as tangled and almost unrecognizable detached masses in saltmarshes where they have been floated by high tides; they may even be found among grasses that can stand inundation with salt water. Such places and the roots of mangrove trees in the tropics are as far as seaweeds adventure on to land. It is said that the laminarians *Nereocystis* and *Macrocystis* can persist as tangled floating masses in the quiet water of sounds. Free-living and sterile masses of the filamentous green alga *Cladophora* are common in the Baltic Sea, where they are rolled into round lumps along the bottom. Recently there was recorded the unique occurrence of immense numbers of floating balls, 1 cm. wide, of the filamentous red alga *Antithamnion* in the sea off Victoria in Australia; 96% of these were sterile, but the remainder had spores characteristic of red seaweeds. There was no evidence where they had come from [23].

The laminarians excel in size and strength yet they grow in a way that seems unsuitable. The young plant forms a stem and a frond, which broadens into the familiar oarweed. At the junction of the two, where maximum strength might be expected, there is the growing region. It adds to the length of the stem and chiefly to the length of the frond, but so subtle is the construction of the core of strengthening tissue that no weakening is engendered by the formation of soft new growth. How different from the leaf of the palm, of comparable magnitude, the soft growing base of which is supported by the mature leaf sheaths! The laminarian frond multiples by means of splits, which arise in this growing base and extending along the frond divide it lengthwise. Then each frond grows basally and splits again, and the daughter fronds continue in the same way. Some laminarians, such as the frilly oarweed (*Laminaria saccharina*), have undivided fronds, and the ribbed oarweeds (*Alaria*), which specialize on headlands, have undivided fronds strengthened by a midrib, as in *Fucus*, but wearing away at the tip as it ages. Others have four to a dozen or more fronds at the top of the stalk.

The giant *Macrocystis* of the Pacific west of North America, the

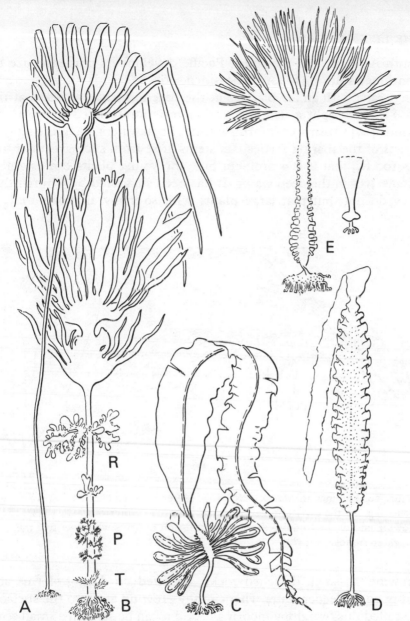

Figure 25. Brown laminarians. A, Bull kelp *Nereocystis*; $\times \frac{1}{40}$. B, Oarweed *Laminaria cloustoni*, growing the new frond below the old one of the previous year, with the red seaweeds *Rhodymenia* (R), *Ptilota* (P), and *Callithamnion* (T) growing on its stem; $\times \frac{1}{16}$. C, Daberlocks *Alaria esculenta*, with its fertile leaflets from the stem; $\times \frac{1}{10}$. (After Newton, 1931.) D, Oarweed *Laminaria saccharina*, growing the new frond below the old one; $\times \frac{1}{10}$. E, Furbelows *Saccorhiza*, with the base of a young plant to show the bulb forming above the first hold-fast; $\times \frac{1}{16}$.

73

south Atlantic, and the south Pacific develops its immense size by continued splitting to form one-sided series of fronds reaching to 4 ft. long, each with an air-bladder at the base to buoy it up, and at the base of each split a new piece of stem is intercalated to space the fronds [24]. *Macrocystis* forms natural breakwaters along the stormy coasts of the roaring forties. Its stems arise from such great depths, 50–100 ft., that it is a problem how the young plants can establish themselves in the deep water. It has been suggested by Church that they do not, but that large plants with so many air-floats may, if

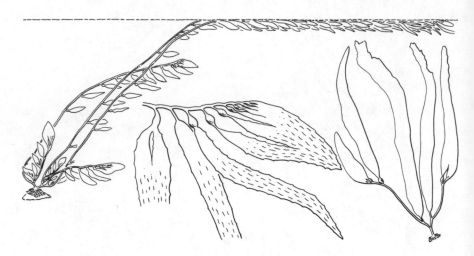

Figure 26. Laminarian *Macrocystis*; $\times \frac{1}{40}$. (After J. D. Hooker, 1847.) A young plant showing the early bifurcation of the stem; $\times \frac{1}{4}$ (after Setchell, 1932); the growing end of the frond to show the succession of splits which set free the daughter-fronds.

growing on small detached rocks, be lifted up during storms and dropped in deeper water whence they grow up again to the surface; he called this 'weighing moorings', and it can be seen on a small scale with the laminarians on British coasts [25]. A more subtle question arises nevertheless in how the strong branching hold-fast of *Macrocystis* can feed.

It seems, as mentioned in the preceding chapter, that there is a downward passage of food material from the fronds to the darkened hold-fast, and that this passage is through the lengthened cells in the core of the stem. The evidence lies not only in the growth of the hold-fast itself, but in the frequent proliferation of new fronds from the base of the stem. The ribbed oarweeds, *Alaria* and its allies, are a particular case, for they develop the fertile fronds bearing zoospores from the short stalk of the main frond, which may be 5–90 ft. long (figure 25C). For what use would all this frond be if the excess of its growth were not conveyed to the lower part of the plant placed under less favourable conditions? It is well known that the stems of laminarians are sweet, as that of *Laminaria saccharina*, and in Japan a thick nutritious soup is made from dried and powdered laminarian stems. They become, as in land plants, the food stores. Many temperate seaweeds, brown and red, overwinter in the cold, ill-lit waters by their stubby basal parts, where the excess of the summer frond, cast in autumn, is stored and whence new fronds proliferate in spring.

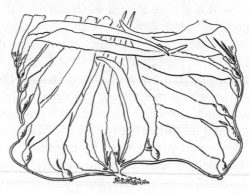

Figure 27. Young plant of *Macrocystis*, showing the development of the long stem by the one-sided growth from the end-frond; × ⅙. (After Setchell, 1932.)

Like other seaweeds, most laminarians subside on to the rocks when the spring tide exposes them for an hour or two. The stalks of some, however, are rigid enough to stay erect, though fronds droop. This is the habit of *Laminaria digitata* and the larger *Laminaria cloustoni* of British shores. It is the habit of the sea palm *Postelsia* of the rough headlands of the Pacific coast of North America; and it is the habit pre-eminently of the palm-like *Lessonia* of southern Chile. The stems are not woody, but the thickened cell-walls are held strongly together and present a mixture of flexibility and rigidity so that they can be snapped. The rigid laminarians have no air-floats, but the rigidity of the stem holds the fronds in the flowing water as crest and trough of the waves of oscillation pass over them. They contrast strongly with the bull kelp *Nereocystis*, which, with the biggest air-bladder of all, trails its fronds near the surface.

Yet a third way of laminarian life is shown by the grotesque furbelows *Saccorhiza* of the Atlantic coast of Europe (figure 25E). From cliff tops its groves may be seen dimly swaying in submarine canyons. Each plant has several oarweed fronds on the top of a flat stem, well adapted to swaying in one direction but rigid in the other; along the rigid edges, where the water flows and eddies, develop the wavy furbelows. The stem is attached to a swollen structure called the bell, which resembles a somewhat flattened football studded on its lower side with attachment roots. Where the stem joins the bell it takes a twist upon itself, thus allowing the stem and fronds to oscillate in any direction on the firmly attached bell. It sways with the oncoming waves and with their reflections from the rocky walls of the canyon.

How does the twist arise? Is it inherited, and will a plant produce a twist if grown in still water, or is it the direct effect of surge on the growing plant? What of the furbelows? Are they inherited or do they develop only in the slipstreams of the flat stem? What of the bell? Is it inherited and will it form in still water or does it need the same external impacts as may induce the twist and furbelows? What of the whole plant form? Is it inherited so that it will appear under any viable conditions, or does it require rhythmically moving water to elicit it? What happens to *Saccorhiza* if grown in a still aquarium? Very small plants have been grown but not to any appreciable size that will answer these questions. They can in still water form a diminutive hold-fast, stem, and frond. Yet there is the remarkable case of *Laminaria saccharina*, normally several feet long and composed of millions of cells worked into an ungainly but efficacious thallus. In the aquarium it can reproduce as will be explained in the next Chapter as a zygote or single attached cell, out of which come zoospores and thus eliminate entirely the large thallus (figure 31). Imagine a sunflower doing this! The sunflower is so complicated and stereotyped that it cannot metamorphose in this way. Like all land plants it is bound by the instructions of a marine inheritance modified to land usage and it will endeavour to become itself under any circumstances. Thus by inspection the hereditary characters of land plants can generally be told. Sea plants grow where they are wrought. They must inherit the means of responding to the actions of the environment but, by inspection, it cannot be told what feature is inherited and what induced by the environment. Remove the natural play of external forces, as tide and surge, and the seaweed form may not materialize. We stumble on a mighty problem.

There are two ways of considering the twist of furbelows. The twist is an adaptation for freely swaying. It might be considered that the motion of the water acting on young plants of the ancestral furbelows caused the twist and, little by little, in the course of generations the tendency to twist had been incorporated into the growth of the plant and become its inherited manner of growth as a property of the chromosomes. This is the theory of the inheritance of acquired characters, which biologists now deny strongly because they have found no evidence for it and because they can see no physical or chemical train of events by which it can happen. Instead, they would suggest that some changes took place in the chemistry of the chromosome that led to twisting: twisting plants would grow better, reproduce more, survive more, and thus would be selected in the struggle for existence by virtue of the internal evolution of the chromosomes. This is the theory of hereditary change by spontaneous chromosomal modification and the natural selection of the modified offspring: the environment chooses but in no way makes the fashion. An opponent of this view might ask by what physical or chemical means could the chromosomes cause the twist without there having been a twist or any previous motivation on the part of the chromosomes. In fact we do not know the answer. When two theories conflict there is often an assuaging middle course. Practically nothing is known of the hereditary mechanism of seaweeds, and they may run the middle course. Chromosomes they have and therefore, by analogy with other organisms, inherited characters but unless the natural environment is there to stimulate practically nothing may result. The extraneous and the innate forces combine in the formative environment of the shore to make the plant.

Consider the rough stalk of *Laminaria cloustoni*. Attached to its end are its own large fronds, which thrive in the surge (figure 25B). They are renewed annually during the seaweed's life of ten to twelve years. High on the stalk, yet sheltered by the fronds, which baffle the water flow, and not subject therefore to the same thrust of the wave as the stalk there grows the red *Rhodymenia*, the fronds of which mimic in their false parenchyma the truly parenchymatous structure of the laminarian. On the base of the stalk, fully sheltered and shaded, grows the delicate filamentous red seaweed *Callithamnion*, in no way compacted into a tough plant body. Along the stalk, from below upwards, grow other red seaweeds with progressively stronger and corticated construction, for example *Ptilota*. Here on one plant is a series that indicates the evolution of the many-celled plant body from the open filamentous, exposing all the filaments for absorption in relatively quiet and therefore rather dimly lit water, through various degrees of compaction (which strengthen the body but do not lessen its absorbing power because the water moves more rapidly) to the most compact, toughest and largest in the free run of the great waves.

All these seaweeds on the *Laminaria* stalk can grow independently also on rocks, mostly in pools, where there are similar environmental conditions, chiefly of light, water movement, and exposure. The slender filaments can grow in the most sheltered places or on the stems and fronds of stouter seaweeds that take up the strain of the water and which may themselves be set with their own filamentous outgrowths or mucilage hairs in which the foreign filaments nestle. Everywhere in the sea, plant building goes with plant action and environmental situation. It is not so on land, where trees grow on trees and the epiphyte has the same structure as the support; where shrubs may grow underground; where mosses are merely mosses and nothing better wherever they occur; and where filamentous hairs come and go without relation to the environment. Botany has been thought out in this perfected light but, as plants have been, so must botany be shaped by the sea.

Chapter 6

Reproduction and Wastage

THE SEAWEED begins to reproduce when it has reached adult form, though not necessarily full size. It sends into the water thousands, and in the larger seaweeds millions, of microscopic reproductive cells. In this way it continues until the whole body is exhausted and from the hold-fast upwards begins to die. Most seaweed bodies have short lives of a few months to a year or two. Some can overwinter in colder waters and live for several years, and in this way the larger oarweeds can endure for ten or twelve years. It seems, also, that the nullipores that cement the coral reefs must be able to survive even longer. But there is little precise information on these matters.

The reproductive cells may be zoospores or gametes, motile with flagella or non-motile, but in either case rejuvenated and without a cellulose wall. They are naked plankton units extruded from the walled cells and, in general, they are remarkably alike, even to the electronic eye. Every species nevertheless has its specific character, or memory, located in the nuclear chromosomes surrounded with the operative cytoplasm in which are one or more of the characteristic chromoplasts. The units are fully equipped to grow into new seaweeds and have therefore a more complicated interior than the plant flagellates. First they swim and drift for some hours or days, such exact details having rarely been discovered; then they attach themselves to the rocks. How they can do this in moving water is difficult to see but it has been shown in some cases that the zoospores swim away from strong light. The reaction brings them to the rocks where they are able to form very quickly the adhesive wall in contact with the hard surface.

The majority of reproductive units will, however, be floated out to sea. There if not eaten by animals they will die because, in spite of their planktonic nature, they have lost the ability to grow in these conditions. Random and wasteful as this manner of reproduction is,

79

the number of reproductive units is so large that enough survive to settle, though their misfortunes are not over. Of the many that start comparatively few reach maturity. Most fail because they have settled in unsuitable places and are overgrown by others adapted to the conditions. Even in suitable places many will be crowded out by the early arrivals, by the fortunate that benefit from some local advantage, and by those which are innately more vigorous. The able that can find their right place, the strong that defeat their kin, and the lucky are the survivors. If indeed the number of adults of a species remains roughly constant on the shore, only one reproductive unit per adult will on the average survive. This fate of attached plants, familiar as the fate of wild seeds and seedlings but dating from the establishment of benthic plants, is merely the costly substitution of the dead and dying. Nevertheless the bigger the adult the greater will be its reproductive output and the greater the chance of survival. Any excess increases the population, and this will lead in due course to the spreading of the species and to geographical distribution. Reproduction, death, dispersal, re-establishment, and distribution are elementary points in biology, but it must be understood how they have entered plant life. To teach from the higher plants of the land may be easy, but it is not profound. Thus Church wrote in *Thalassiophyta*: 'the recognition of the great "Law of Benthic Waste" is the beginning of wisdom', and 'the Benthic Phase introduces . . . . the Benthic Law under which the individual exists solely for the good of the race, and the race is forwarded at the individual expense' [26].

In lowly seaweeds of filamentous or plate-like form, all the cells are much alike except the attachment cell, and any cell other than this one may reproduce (figures 12, 13). As seaweeds become more complicated by branching, thickening, and cell-differentiation, the general ability is lost and only certain cells, or groups of cells, maintain the power to reproduce. Such a cell is a mother-cell for the reproductive units; if these are zoospores or asexual spores it is called a sporangium; if they are gametes it is called a gametangium. In the larger filamentous seaweeds certain end cells or intercalary cells or groups of cells on short branches are often said to be specialized for reproduction but, in fact, it is these cells that retain the primitive capacity and are not sterilized into the vegetative state as befalls the majority. In larger seaweeds the variety increases. Patches of surface cells may be reproductive, or groups of more or less heterotrophic internal cells, and such patches or internal groups may be restricted to special branches and fronds shorter than the normal vegetative ones because of their limited life. Such a cluster of reproductive cells is called a sorus.

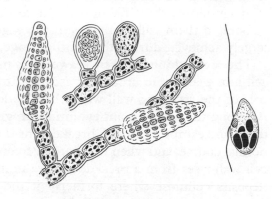

Figure 28. Parts of the brown filamentous seaweed *Ectocarpus* with gametangia (pointed, plurilocular) and sporangia (rounded, unilocular); × 500. A zoospore with eye-spot, chromoplasts, and two unequal flagella attached laterally; × 2000.

The plant body becomes divided therefore not only into special vegetative parts with special internal structures but also into vegetative and reproductive cells, and thereby into long-lived vegetative parts and short-lived reproductive parts that may have special sterile or vegetative tissue attendant on them. Though they have nothing comparable with the inflorescences of flowering plants in which this distinction is exaggerated, the seaweeds begin the division of labour into the sterile vegetative body, the surplus material of which is later concentrated into the reproductive organs, and they show how the reproductive parts become themselves further differentiated and elaborated.

The greater the division of labour the deeper is our insight because the elaboration renders the nature of the plant apparent. Vegetative differentiation is very generalized and largely common property, but reproductive differentiation extending from the flagellate characters of the zoospores and gametes upwards to those of sporangium, gametangium, and sorus become more and more distinctive. The reproductive features become in fact the main criteria for the classification of seaweeds into genera, families, and higher ranks. Some of these features are also closely related with the environment, in which case they occur in parallel in the green, brown, and red seaweeds. Others are more peculiar and seem to reflect the intrinsic nature of that kind of seaweed. Thus it is that the classification of all plants comes to be based chiefly on the methods of reproduction that they

81

have inherited, and too often sight is lost of the vegetative parts which necessarily precede the reproductive and have enabled their evolution. Benthic seaweeds inherited their methods of reproduction from their planktonic ancestors and elaborated them greatly; land plants inherited these elaborations, according to their own kinds, and elaborated them still further, and the vegetative differentiation was largely achieved during the benthic stage.

The method inherited by seaweeds is multiple fission of a mother cell into planktonic units that escape through a pore or slit in that part of the retaining wall which breaks down. These verbs are apt. All along the line of plant evolution progress has come from breakdown and escape. Through breakdowns there are deviations from the normal course, and when these are overcome there is a new way. The cell-wall rises from a breakdown in the life of a naked flagellate; it deposits cellulose where formerly it did not, and the use of this deposit leads to the plant-cell. Its planktonic existence breaks down, and it sinks to become the benthic plant body which, if it did not break down, would not liberate the planktonic units for reproduction. If internal cells break down, the thallus is weakened with mucilage or buoyed up with air-bladders; if the breakdown is repaired by strengthening the walls and lengthening the cells, the internal cable of the robust seaweed is formed. The problem before the student of evolution is to consider what may have gone wrong with the intricate living mechanism to have altered it; if nothing goes wrong, it is perfect and will continue unaltered, but mistakes can lead to success.

On liberation the planktonic units behave in one of two ways. They are either asexual zoospores that develop directly into the seaweed or they are gametes which conjugate and the zygotes settle to become the seaweed. In this case the seaweed is diploid: somewhere in its life-cycle meiosis must occur. A direct solution would be to consider that the seaweed had two states; one would be diploid and produce by meiosis the zoospores that would grow into the haploid plant, and this would produce gametes without meiosis; thus the diploid zygote would start the life-cycle again. Many seaweeds have this double state. A familiar example is the sea lettuce *Ulva*. Its two states appear identical, but one is diploid and forms zoospores from any of its cells (except the internal hyphae and the hold-fast), and the other is haploid and forms similarly the gametes. Because it produces the asexual spores, the diploid plant is called the sporophyte;

the haploid plant is accordingly the gametophyte. These names have to be learnt; they figure repeatedly in the life-histories of many-celled plants, and the lives of all successful land plants are built on this scheme of a direct alternation of gametophyte and sporophyte.

Alternation of generations, sexual with asexual, is a familiar phrase in botany. The two generations may be exactly alike, as they are in *Ulva*, and differ merely in chromosome number and sexuality; this is called isomorphic alternation. In contrast they may differ more or less strongly in size, form, and manner of growth, in which case they will occupy different positions in marine vegetation, and this is the heteromorphic alternation so marked, as will be explained, in the laminarians and so indicative of their superiority. In higher land plants, such as the seed plants, the sexual generation or gametophyte becomes almost suppressed, for it is clearly not feasible for the successful colonizers of dry land to be dependent on a supply of standing water into which to discharge free-living gametes. In the sea there is no such overriding problem and therefore no such uniformity in the life-cycles that have been exploited in all manner of variations in the colonization of the shore.

The ideally direct alternation is shown schematically by the double lines in figure 29. In some brown seaweeds a detail assists in identifying the two generations. The zoospores are usually formed in the cavity of the mother-cell, which is called a unilocular sporangium, whereas the gametes are formed by the division of the enlarged mother cell into many small units all with their own walls, and from every one emerges a single gamete: such are called plurilocular sporangia or, strictly, the gametangia. Diploid plants then have unilocular sporangia and haploid, sexual plants the plurilocular. However, as shown by the broken lines in figure 29, there are exceptions. The haploid gametes may not unite in pairs, but behave as zoospores and develop into another haploid generation and, so long as there is no conjunction, this haploid sporophytic (asexual) state will continue. Alternatively, meiosis may fail in the unilocular sporangia so that diploid zoospores are produced that never conjugate but continue a line of diploid sporophytes. In this case both unilocular (presumably meiotic) and plurilocular (mitotic) sporangia can be found on the same plant. Such variations, dependent apparently on external conditions, can greatly complicate the life-histories of the simpler seaweeds in which, one supposes, the ideal rhythm has not been fixed. An eminent example is that of the brown filamentous

Ectocarpus and its allies, the life-cycles of which may vary according to the geographical region in which they grow.

In other cases, of which both the brown fucoids and the green *Caulerpa* and *Codium* are examples, there is no alternation of generations (as shown under *Fucus* in figure 29). The plant body forms only gametes, preceded in their unilocular gametangia by meiosis. Thus

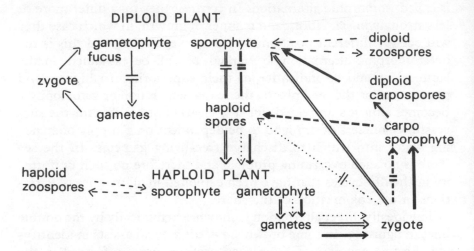

Figure 29. Life-cycles of seaweeds. Paired cross-lines mark the position of meiosis. Paired lines show the cycle of the laminarians and that of mosses, ferns, and seed plants. Dotted lines show the simple cycle of many green seaweeds and plankton cells with meiosis at the first intent, without the diploid plant stage. Thin lines show the cycle of the fucoids without a haploid plant stage; meiosis at the second intent. Broken lines show the variations in many filamentous green and brown seaweeds. Thick lines show the cycle of red seaweeds with meiosis at the third intent; thick broken lines show the cycle of those without the free-living sporophyte with tetraspores. Note that sporophyte and gametophyte may be either haploid or diploid.

the plant is diploid in chromosome number but gametophytic, just as in the normal condition in animals. In still other cases (shown by the dotted line in figure 29), principally among the filamentous green kinds such as *Ulothrix*, and the yellow–green, meiosis always occurs, as in *Chlamydomonas*, at the first division of the zygote, which liberates haploid zoospores. These may continue a line of plants that produce zoospores and function as asexual and haploid sporophytes or they may produce gametes and function as gametophytes; they have no diploid plant form. Whereas it is possible to generalize about land plants such as mosses, ferns, and seed plants, that the diploid plant is asexual and, rightly, sporophytic, and that the haploid plant is sexual and, rightly, gametophytic, no such generalization can be made about the seaweeds, which, in their experimental state, may have diploid gametophytes and haploid sporophytes as well as the two states normal for green land plants.

The simplest form of sexual reproduction is the union of two flagellate gametes that in all appearances are alike. They also resemble the asexual zoospores of the plant, and this adds to the difficulty of working out the life-cycles of the simpler seaweeds. In some, however, the diploid zoospore is larger than the haploid gamete, and in others such as *Ulva* (figures 12, 13) and the filamentous green *Cladophora*, the zoospore has four flagella, the gamete two. Gametes similar in appearance must nevertheless, as mentioned in Chapter 2, differ in sexuality. It is known that in some filamentous seaweeds those produced by one individual will not pair together but only with those from another individual. These individuals appear as alike as the gametes yet one must be inherently male and the other female. Since no chemical test is known to distinguish such incipient sexuality, the custom is to designate the individuals by the over-worked symbols + and −. Sexual reproduction is then accomplished by the union of a + gamete with a − gamete, and the process is called isogamy or the union of outwardly similar gametes.

As mentioned also in Chapter 2, flagellate gametes may be attracted to each other by two means. All may be attracted to light of a certain intensity and thus be drawn together. The red eye-spot detects light, but it is improbable that this simple structure, even with a lens, enables the gametes actually to see each other by means of an optical image. Then one gamete may taste and be attracted by the excretion of the other and this, in its simplest form, would mean that a chemical excreted by a gamete of one sex would affect flagellar activity in the other so as to direct it towards the source of increasing concentration of the chemical. It is known that the male gametes of some mosses and ferns are attracted to certain concentrations of simple organic acids. Root-hairs, pollen-tubes, fungus hyphae, and other land filaments may also show such directed growth. Taste therefore, in its simplest form, must be accepted as a faculty of protoplasm, whether plant or animal, and a cause of selective mating.

85

If two persons wish to find each other it is better for one to wait while the other searches. And if on meeting they must journey, it is better if she who waits should be provisioned while he that searches may travel light and fast. The principle of assignation was worked out long ago by gametes. A well-stocked egg, without flagella and too big to be moved by flagella, is fertilized by a small or undersized but active male flagellate as the spermatozoon. The evolutionary process is finalized when the egg-cell is retained within its parent organ (gametangium) on the parent plant and fertilized *in situ*. Red sea-weeds have come to this conclusion and so, almost, have the laminarians. It is the way of all higher green plants of the land and of many of the freshwater green Chlorophyceae, though not their modern green seaweeds.

Heterogamy is the name given to the conjugation of dissimilar gametes. If they are sperms and eggs their conjugation is called oogamy; the male gametangium is then called the antheridium (antherozoid being a botanical alternative to spermatozoid) and the female is called the oogonium. Reproduction by eggs has arisen presumably through the selection and inheritance of variations that have helped both the mating of the gametes and the establishment of the zygote as a new plant. The variations have led to the breakdown of female mobility, but out of this has come better provision for the offspring.

A number of seaweeds, even *Chlamydomonas* among planktonic cells, show very remarkably the steps in the consummation of oogamy that might be expected. In some, for instance, a flagellate gamete soon tires and, on stopping, is fertilized by a male of the same size but more active. In others, such as the green *Codium*, the less active gamete is the larger. Then in *Fucus* the large non-motile egg is extruded and fertilized in the sea water. In *Laminaria* the extruded egg sits on the slight filament that produced it and is fertilized practically *in situ*. In the red seaweeds the egg-cell never escapes from its oogonium, where the zygote is nourished by the parent plant. Thus maleness and femaleness can be seen to differentiate from the isogamous beginning. Among animals there generally goes with the gamete distinction maleness and femaleness of the animal body or individual animal, but among plants this addition is not of primary importance and need not happen. One and the same plant body commonly produces both sperms and eggs, though it may not follow that sperms can fertilize eggs of the same parentage. This common state is called monoecious (in one house). The alternative with male and female plant bodies is called dioecious (in two houses); it is comparatively rare and always needs emphasis. The remarkable point about these slight differences in gamete behaviour, however, is that there should still be plants of relative simplicity satisfied with what seem imperfect, as they are intermediate, steps in the evolution

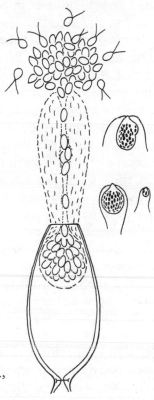

Figure 30. Gametangium of the green seaweed *Codium*, opened and extruding the gametes in mucilage; ×200. The union of the motile gametes of unequal size; ×300. (After Smith, G., 1938.)

of reproduction by eggs, whereas most plants have long ago completed this lap of evolution's race. There are two explanations.

Firstly, just as the isogamous seaweeds have not progressed at all in respect of sexual evolution, so the intermediate steps shown by *Codium* and *Laminaria* may have been reached long ago and have, ever since, been stationary. Secondly, *Codium* and *Laminaria* may be in the process of sexual evolution on their own account, *Codium* being the more recent as it is the less advanced. Whereas the main course of plant evolution has certainly been run, there are plants nevertheless that start late and have not finished, though what they may achieve will be merely a repetition of what others have done. The lesson is that the intermediate states, so fortunate for biological understanding, may seem small, even trifling, yet they may fit well environmental conditions that, as regards microscopic life in particular, we may be unable to appreciate. The subject could be extended greatly, as ever in biology, for when the student has become attuned to the incessant questioning how, why, where, and when, in the words of the poet Blake 'one thought fills immensity'.

The heteromorphic life-cycle of laminarians contrasts strongly with that of *Ulva*. It is a strict alternation of generations in which the large oarweed is the diploid sporophyte but the haploid gametophyte is a microscopic creeping filament comparable with the very early filamentous stage of *Ulva* before it expands into the plate of cells. The largest seaweeds therefore are sporophytes and they change in the course of their lives into the smallest of gametophytes. To detect them the young laminarian plants must be traced back to their microscopic origin; the zoospores must be grown on microscope slides in aquaria or fixed to the rocks beneath the oarweeds; and most botanists have never seen the living gametophytes.

The sporangia of *Laminaria* are borne in dark patches, or sori, on the surface of the frond. In the ribbed *Alaria* the sori are on the short and thicker fronds that arise from either side of the stalk. The haploid zoospores are liberated in millions, and the survivors grow into the short, sparingly branched, creeping, and minute filament on which several antheridia and one or two oogonia are borne. The sperms have the form characteristic of brown seaweeds: the pear-shaped body has two unequal flagella inserted on the side, as in the peridinean, with one directed forwards and the other backwards. They swim to the eggs that have been extruded from the oogonia, one from each, and wait on the outside of the empty walls. The zygote is not detached but grows where it was made and soon comes to smother the gametophyte; by January or February, after autumn fertilization, the small brown blades of the sporophytes become visible to the naked eye. The gametophytes of *Saccorhiza* may be dioecious and the female may consist merely of an attachment cell and the oogonium or of a single cell that serves both purposes; it is thus a minimal many-celled plant.

This alpha and omega of plant construction fits the environmental extremes. The large tough sporophytes grow in the best situations and mobilize in consequence a maximum number of zoospores that reproduce, multiply, and disperse the species: they meet the enormous wastage necessary for re-establishment on the shore. The minute gametophytes grow freely in ill-lit crevices among the holdfasts of the sporophyte, where water movement is minimized, and they carry out the function of sexual reproduction with minimum

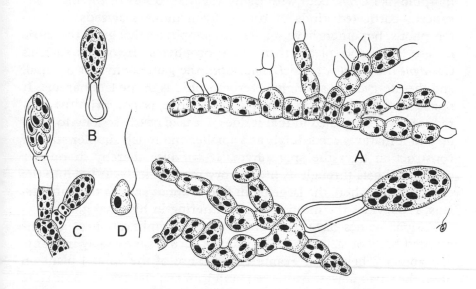

Figure 31. Microscopic filamentous life of the laminarians. A, Gametophyte of *Laminaria digitata*: male (above) with empty antheridia, female (below) with the egg extruded from the oogonium; ×900. (After Kylin, 1916.) B, Female gametophyte of *Saccorhiza* limited to one cell, which has become the oogonium and extruded the egg; ×900. (After Sauvageau, 1916.) C, Female gametophyte of *Laminaria saccharina* with the fertilized egg transformed into a sporangium with zoospores; ×900. (After Pascher, 1918.) D, Sperm of *Alaria*; ×1200.

wastage and that among the sperms only. The gametophyte employs the primitive environment of benthic plants, the sporophyte the most advanced. In other words, heteromorphic alternation is the way the two states of the seaweed exploit and explain the environment. Laminarians have a small gametophyte but other brown seaweeds, such as *Cutleria*, have a small sporophyte; its two states were classified as different genera until Church proved that they belonged to the same species. When therefore the laminarian type of life-cycle is found to be characteristic also of the dominant land plants, namely ferns and seed plants, it is at once understood as part of their marine heritage, which, by virtue of the small gametophyte, enabled them to overcome the problem of free-living gametes on land. Brown seaweeds must tell the story because the green seaweed ancestors of the land plants have disappeared.

Steps exist, too, that explain this life-cycle. *Ectocarpus* is isomorphic like *Ulva*, both with many vagaries. Some of the more advanced, corticated kinds of brown filamentous seaweeds are isomorphous, but others have smaller gametophytes that grow therefore in less favourable places than the sporophytes. In others such as *Castagnea*, *Mesogloia*, and *Desmarestia*, the gametophyte is as small and the life-cycle as strongly heteromorphic as in the laminarians. It seems therefore that the small gametophyte is not primitive but reduced and simplified in that it does not take nearly so long to grow up and it matures accordingly at a smaller and much simpler state of construction than the sporophyte. It survives thereby in quieter waters and gets through its life-history more quickly. It fits into the environment where the large plant body cannot and, in the laminarians, it provides the means of overwintering in hibernal gloom.

Various names have been given to this simplification of a complicated body or structure. The process occurs very widely in plants and animals, being, for example, the cause of the annual plant on land, the small flower, fruit, seed, and leaf, and the microscopic state of many fungi. Neoteny (attaining in youth), paedogenesis (childhood reproduction), juvenescence (becoming young), or, plainly, reduction (leading back) describe the way in which lengthy development is cut down and a secondarily simple, or simplified, plant or animal derives what Church called 'quick returns'. Nowadays, with plant evolution fulfilled, the simple are frequently the simplified, not the primitive. That is why it is necessary to understand the main sequence and environmental instigation of plant evolution and to be able to detect the backsliding and the short cuts.

These thoughts lead to *Fucus*. There are no zoospores in the fucoids and therefore they have no alternation of generations. The diploid plant produces sperms and large non-motile eggs, as distinct from each other as in most animals. They must employ the environment therefore very differently from the laminarians. The eggs in oogonia and the sperms in antheridia are set in small mucilage depressions, called conceptacles, which are themselves generally

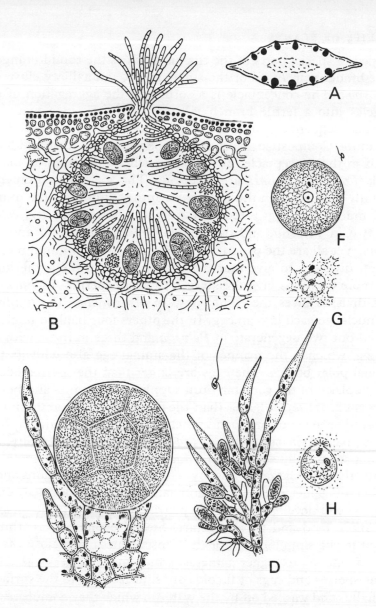

Figure 32. Reproduction of *Fucus*. A, Cross-section of the fertile tip of a branch to show the conceptacles; ×2. B, Conceptacle with oogonia and clusters of small antheridia, with mucilage hairs (paraphyses) projecting into the cavity of the conceptacle; ×40. C, Oogonium with eight eggs, and D, a mucilage hair with antheridia on its branches; ×160. F, Fertilized egg with a sperm inside it making its way to the female nucleus; ×200. G, Small sperm nucleus uniting with the large egg nucleus; ×500. H, Sperm nucleus forming chromosomes to pair with those of the egg nucleus; ×1000. (A–D after Thuret, 1878; F after Farmer, 1898; G–H after Yamanouchi, 1909.)

grouped in special parts of the thallus, such as the tumid orange tip of the branches in *Fucus* or the small, spiky, and axillary clusters in *Sargassum*. The conceptacle is a sorus and the aggregation of conceptacles into a fertile branch is a sorus of sori, in a manner of synthesis frequent throughout the plant kingdom even to the flower head of the Compositae. Most are monoecious, with male and female organs in the conceptacle, but others, such as the common bladder wrack (*Fucus vesiculosus*), are dioecious. The sperms are numerous, generally sixty-four, in the unilocular antheridium, but the number of the much larger eggs is reduced to eight in *Fucus*, four in *Ascophyllum*, two in *Pelvetia*, and one in *Halidrys*, *Himanthalia* and *Sargassum*. Yet all are the products of meiosis with its minimum of two nuclear divisions to give four daughter-nuclei. In the antheridium, four more divisions give the sixty-four small sperms. In *Fucus* one more division gives the eight eggs. In *Ascophyllum* the minimum four nuclei go each into an egg. In the others four haploid nuclei are formed but two degenerate in *Pelvetia* and three in those with only one egg, which is the manner of the animal egg also with its three vestigial polar bodies. The eggs are larger than the sperms because the cytoplasm of the gametangium supplies fewer units and because the oogonia are much larger than the antheridia. The antheridia are produced at the ends of much branched, if minute, filaments; hence with a fixed food supply they are small as they are many. The oogonia are usually solitary on a filament and, with the same food supply, therefore larger. Thus, by reducing nuclear divisions and the number of gametangia, fewer and larger eggs are formed, and if fertilized and developed they will provide the microscopic offspring with more food and a better start. Sexual evolution can thus be related to the simplification of the consequences of meiosis; failure to subdivide led to bigger gametes, which made better eggs.

The sperms and eggs of fucoids are squeezed out to the surface of the thallus and washed off by the water in which they conjugate. The wastage of reproductive units falls therefore on the zygotes in their chance establishment, and this seems extravagant. Not much is known about the process but in some, certainly, the wastage is reduced by timing the discharge of gametes with the low-tide. Then with fertilization in the ripples and slight surges of the rising tide the zygotes are wafted on to the rocks, where they at once attach themselves: lacking flagella, they cannot make their own way in the manner of zoospores, but within the tide limits the zygotes may enjoy slight dispersal with ready opportunity to fix themselves. If the zygote of the laminarian were set free, it would drift off and undo the successful establishment of the zoospore. The fucoids use the tides for their reproduction, unlike the laminarians in deeper water. Thus in reproduction, as in details of structure, these two groups of parenchymatous seaweeds present profound differences, which suggest that they are parallel attainments, fitting parallel situations on the shore, in consequence of the fundamental difference in life-cycle. In recent books the attempt is made to derive the fucoids from the laminarians by total loss of zoospores and gametophyte, the gametangia of the fucoids substituting the sporangia of the laminarians on the diploid plant; but the suggestion does not explain the comparable life-cycles of *Codium* and *Caulerpa* or the animals, which can hardly be supposed to have lost asexual zoospores and gametophytes. To the marine biologist, unworried by the land flora, the fucoids are a successful line of plants distinguished by a life-cycle similar to that of animals.

Zoospores, gametes, and eggs! If the large size of the egg helps the zygote to start on its new career, why not large zoospores? They could not swim, but the fucoid zygotes show that passive dispersal can be satisfactory and that swimming can be cut out as old-fashioned. The next step in the sea improvement of reproduction seems to have been the introduction of the large non-motile asexual spore, which prevails throughout the red seaweeds as one of their successes in speeding up sporeling growth by means of a bigger food supply. The new habit occurs in the rather slender brown seaweeds, of which the common *Dictyota* is a good example. They are thinly parenchymatous plants of the tide range, thriving on the warmer temperate and the tropical shores where conditions are not too rough. They have apical growth as in the fucoids, and *Dictyota* is like a thin, slender, equally dichotomous *Fucus*: the more tropical *Padina* has the form of a brown fan, generally obscured by numerous proliferations from the tapered stalk. Their life-cycle is an isomorphic alternation of diploid sporophyte with haploid gametophyte, which is also dioecious so that these seaweeds appear in three superficially similar forms, namely the asexual sporophyte and the male and the female gametophyte. Having no speciality in growth, the three forms compete in frequent turnover and occupy no prominent position in the vegetation. They form sperms and eggs that are set free in the water where conjugation occurs. No meiosis initiates the formation

93

of the gametes because the whole gametophyte is haploid, yet the oogonium has advanced to the state of producing only one large egg. The sporangia on the sporophyte produce not many small zoospores but four very large non-motile spores, superficially like the eggs. Four is the sign of meiosis and these four spores, or tetraspores, are haploid. Thus in the sporangium of *Dictyota* and its allies there must have occurred the same reduction in the number of reproductive

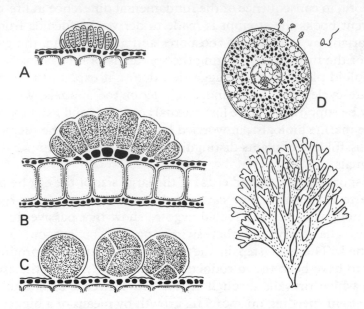

Figure 33. Brown parenchymatous seaweed *Dictyota*; × ⅓. Cluster of antheridia (A), oogonia (B), and tetrasporangia (C), seated on the surface of the frond, which is composed of three layers of cells, the outer small and photo-synthetic (in black) and the inner enlarged; × 120. D, Extruded egg in the sea water with sperms (one flagellum each); × 600. (A–C after Thuret, 1878; D after Williams, 1898.)

units as in the oogonium of the fucoids, but it is not carried on to the state of one spore. Reduction to unity is a female contribution to sexual evolution. The tetraspore, being asexual, seems not to have come under this influence; as a means of dispersal moreover its numbers must be maintained. However, among all the plants that one would least suspect of tricks, the sedges (Cyperaceae) do this very thing; only one of the four pollen-grains, formed in tetrads in the stamen, survives. So nature snaps her green fingers at attempts to rationalize plant behaviour.

The *Dictyota* propagate therefore by means of two generations of egg-like reproductive units, namely the sexual zygote and the asexual tetraspore. The fact is highly significant because tetraspores are the standard spores of mosses, ferns, and seed plants; they are the advanced, well-stocked improvement of the zoospore adaptable to land conditions. *Dictyota* is therefore advanced in this way compared with *Laminaria*, but it has insufficient bodily construction to become a major success on the shore; if it had the laminarian body, its large spores might not be numerous enough to overcome the wastage of re-establishment.

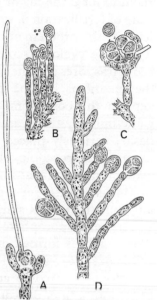

Figure 34. Reproductive organs in the filamentous red seaweed *Ptilothamnion*. A, Oogonium with long trichogyne; ×200. B, Antheridia clustered at the tips of short branches, and the small round spermatia; ×120. C, Cystocarp developing carpospores on diploid filaments from the zygote and surrounded by short haploid filaments from the parent plant, and the remains of the trichogyne; ×120. D, Diploid filament with the tetrasporangia; ×120. (After Bornet, 1867.)

The common *Dictyota dichotoma* has been studied enough to show the more detailed complexity of seaweed behaviour. Around the British Isles the gametophytes form the sexual organs at fortnightly intervals from July onwards. The organs are begun at the neap tides and the gametes are set free before the spring tides. The rhythm is maintained even in plants transferred to an aquarium and grown without tidal effects. The same rhythm has been found at Naples where the tide is extremely slight. In Jamaica, with irregular tides, there seems to be no reproductive rhythm. On the coast of North Carolina, however, there is a monthly rhythm; sex organs are started at the spring tides of full moon and the gametes are set free at the next. Even on a life-cycle so advanced, unpredictable variations may be wrought. Truly, knowledge of seaweeds is very far from perfect.

Red seaweeds are often regarded as too peculiar for ordinary discussion and outside the pale of ordinary botany. But we are interested in plants, and this interest cannot be stopped, particularly when we perceive that the full force of the marine state of plants cannot be appreciated without the red seaweeds. Reproductively they are the most advanced seaweeds and therefore they occur in greatest variety on all coasts. Their bodily construction is not strong enough for them to form the major vegetation, but they have arrived at the final state of fertilization *in situ* when the egg-cell is not liberated; and this leads to the problem of how to get the new generation off the parent. Their life-cycles resemble that of *Dictyota*. The sporophyte bears red tetraspores and the gametophyte (monoecious or dioecious) bears the sex organs, but the sperms are also without flagella and the egg is kept in the oogonium. Then on the gametophyte, or the female gametophyte, there are found deep-red spots, knobs, or minute flasks, variously placed on the branches or fronds. These are the extra, if microscopic, stage in the life-cycle that enables the new generation to escape: this stage is called the carposporophyte, and it produces an extra set of spores called the carpospores. The names are long but the idea is simple; they could be called quit-spores.

The carposporophyte has grown from the zygote and it is diploid. The zygote is formed on the parent gametophyte because the egg-cell is never extruded. The contents of the oogonium on the haploid gametophyte become entirely the egg-cell. From the apex of the oogonium a fine, delicate mucilage hair is produced that extends in microscopic length to the surface or exterior, because the oogonium is a particular cell in a filament that may be surrounded by other filaments and even immersed in the tissue of the plant. On to this hair sticks the sperm, which has been carried passively in the sea water. The sperm absorbs the wall of the hair where it contacts it, and the male nucleus passes down the inside of the hair to unite with the female and initiate the zygote. From the zygote, retained within the oogonium, diploid filaments grow out that form the diploid carpospores singly at their ends. Around the diploid filaments there may grow up haploid and sterile filaments from the gametophyte to form a sterile protective wall. From this minute carposporophyte the carpospores are extruded to be dispersed passively in the sea water, for they have no flagella, to settle down and to grow into diploid sporophytes that form the tetraspores by meiosis. And these on dispersal develop as in *Dictyota* into the gametophytes, which repeat the life-cycle.

Many are the complications of red seaweeds. They carry long names, apt to confuse, yet some must be mentioned by way of illustration. The point is to see what the red seaweeds have accomplished. The oogonium is transformed into an unusual shape and through the new process of fertilization *in situ* it bears the new carposporophyte; hence the oogonium is called the carpogonium and its hair the trichogyne. The sperms have no flagella; few encounter the female hair of their own species and most are wafted off to die. They are small round colourless bodies, $3-5\mu$ wide, produced in vast numbers from certain filaments of the gametophytes; because of their unusual character they are called spermatia. By analogy with *Dictyota* it is assumed that the tetraspores have been evolved from motile zoospores and the spermatia and undifferentiated egg from motile gametes, but as all traces of flagella have been lost, the flagellate ancestry of red seaweeds can only be surmised. Then the carposporophyte may also induce an investment of sterile, haploid filaments from the gametophyte, as another consequence of fertilization, and this double structure is called the cystocarp [27].

Because of their multifarious life-cycle, with its two sets of asexual and egg-like spores discharged into the water, and the variety of their filamentous construction, the red seaweeds insinuate themselves into most parts of the rocky coast. They differ in a great amount of detail both in the bodily structure and form and in the precise positions of the reproductive organs; hence they are presented botanically in many intricate genera and species. Their chief interest lies, however, in the way they have paralleled the other seaweeds and yet have exceeded them in some respects that throw light on certain peculiarities of land vegetation not present in other seaweeds. In their red way they prove that the properties of plants have been evolved through their benthic life as thalassiophytes. The possession of tetraspores, so paramount in land plants, is an example. So also is fertilization *in situ*. The laminarians have nearly reached this stage, but they have what the red seaweed lacks, namely the heteromorphic distinction of sporophyte and gametophyte. If the red seaweed had such a creeping filamentous gametophyte, there would be no need for the carposporophyte, because the sporophyte could establish itself simply by growing over it as the young oarweed does. But it has not, and the carposporophyte is the unique means of getting off the gametophyte, paralleled only by certain fungi of the land.

In a sense the carposporophyte is parasitic. It is photosynthetic, but it holds on to the parent gametophyte and it absorbs some food from it. Thus before the zygote forms the carpospore filaments, it commonly fuses with one or more neighbouring cells, which give up to it most of their cytoplasm, though not their nuclei. These cells are called auxiliary cells (figure 35), and their number and position are again important features of the classification of red seaweeds. In some cases the zygote, not content with mopping up some neighbouring auxiliary cells, sends out filaments that may grow for quite long distances through the tissue of the gametophyte and mop up more auxiliary cells, around which they form additional carpospore-producing filaments or cystocarps. Therefore retention of the zygote on the parent involves the saving of female gametes, the succour of the young by the parent, and a parasitic element in the nature of the embryo.

Parasitic seaweeds are rare. Many small seaweeds grow epiphytically on others but they do not penetrate the tissues of the supporting plant and derive organic nourishment therefrom in the manner of mistletoes. Seaweeds seem remarkably antibiotic, or they are not prone to relinquish their photosynthetic independence. Nevertheless there are a few parasites, notable among which is the brown *Notheia*, which on Australian coasts grows from the conceptacles of the fucoid *Hormosira*, and some of the red seaweeds, all of which possess, however, in their embryos the parasitic tendency. Therefore it is noteworthy that among the fungi of the land, many of which are parasitic, the Ascomycetes (Chapter 13) have main features of structure and reproduction reminiscent of red algae and, even, suggestive of the ancestors of red algae. Littorally and morally the red seaweeds form the underworld of marine vegetation.

A parasite, like an epiphyte, is both smaller than the plant on which it grows and less productive. It can take only what the support provides. If the zygote of the red seaweed became a sporophyte with tetraspores, it would render its one dispersal state inferior and less productive. The carposporophyte appears again as the small intermediary that restores the plant on its free course. Now some red seaweeds, of which *Nemalion* is an example, have no tetrasporophyte; the carpospores grow into the free-living gametophytes. This is a

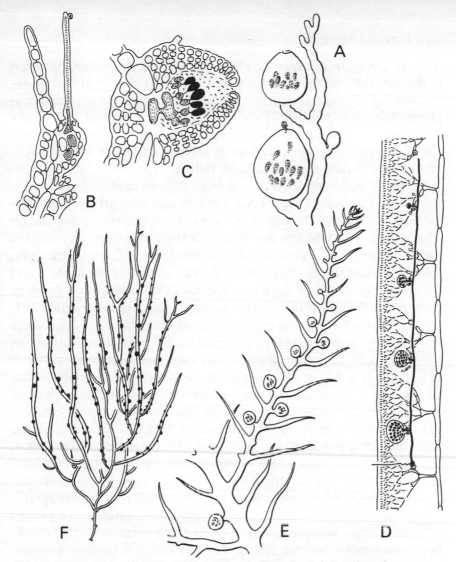

Figure 35. Reproduction of red seaweeds. A, Branch of the corticated
filamentous *Rhodomela* with cystocarps in which the carpospores are forming;
×30. (After Newton, 1931.) B, Oogonium of *Rhodomela* with the trichogyne
and a male cell attached, and with two auxiliary cells (hatched) below the
oogonium; ×250. (After Kylin, 1922.) C, Cystocarp of *Rhodomela* consisting of
a wall of haploid filaments from the parent plant around the diploid carpospore
filaments derived from the zygote united with the auxiliary cells; ×250.
(After Kylin, 1922.) D, Extensive carposporophyte of *Platoma* developed from
the fertilized egg with its trichogyne by means of a long diploid filament
extending to auxiliary cells on successive lateral filaments of the thallus and
forming a cluster of carpospores (in black) at each union. (After Kuckuck,
1912.) E, Part of the plant of *Bonnemaisonia*, showing the cystocarps alternating
on the short branch of successive pairs of branches; ×10. F, *Gracilaria* with
cystocarps; ×½.

simpler life-cycle and may be regarded as more primitive, but such is the intricacy of the details of red seaweeds that evidence is accumulating to suggest rather that this state without a tetrasporophyte, though *a priori* unlikely, is in fact the result of loss of the tetrasporophyte coupled with a return to meiosis in the zygote. There is still a very great deal to be discovered before a satisfactory history of the red seaweeds can be written.

Plant life in the sea has evolved from microscopic size and virtual immortality to mortal plant bodies of size and strength commensurate with the waves. Thereby they have created shelters under and on which lesser plant bodies survive. Into these conditions have crept the diversity in life-cycles, the differentiation of sex, tetraspory, parasitism, and the assertion of the diploid state over the haploid and of the sporophytic over the gametophytic. The trends have occurred independently in the three main series of seaweeds, by which it is meant that each series has met and solved the problems of the shore with similar result; none has borrowed by cross-fertilization the hereditary characters of another, but each has developed its own. Any plant if it was to succeed must have evolved through similar stages, but some went one way, others another, and none was so successful that it suited all conditions. The next great problem is to discover what properties of the marine heritage have led to the colonization of the land, and how land plants have used them. But, lest there be any doubt of the obligate nature of this plant evolution in the sea, brief mention must be made of the blue–green seaweeds.

The blue–green seaweeds (Cyanophyceae) offer the most remarkable parallel in body-building. They imitate in single cells, groups of cells, plates, filaments, and clustered filaments both the planktonic and simpler benthic plants. Yet they differ so fundamentally in cell contents as to suggest a lower state of cell organization than that at which the ordinary plant flagellate or plant-cell starts. Their cells have no such precise nucleus, chromoplasts, watery vacuoles, flagella, zoospores, or gametes, or any method of reproduction other than growing, dividing, and loosening cells or groups of cells to be dispersed by water currents. Their peripheral protoplasm contains chlorophyll, phycoerythrin (very similar to that of the red seaweeds), and the blue–green pigment phycocyanin, which usually preponderates, but how the pigments are held in the protoplasm is not clear. They are the least understood of plants but their turn is coming with the electron-microscope because they connect with bacteria.

It seems, in fact, that they are always associated with bacteria that live on the mucilage of the blue–green cells; if, in the laboratory, they can be freed of these bacteria then the blue–green cells grow poorly. Some tend to live on decaying organic matter in the manner of bacteria, in which case they may lose their photosynthetic pigments and become, if small, so like single bacterial cells or bacterial filaments that it is not possible to say how they differ. Whether bacteria came from blue–green seaweeds or vice versa, and whether either is connected with the previous history of green, brown, and red plants are open questions [28].

Appendix

Some may wish to master the problem of seaweed life-cycles. There is much research needed with an open and inquisitive mind. I offer the following guidance, based on Church's understanding, because it is the only comprehensive attempt. The problem centres on conjugation and meiosis, and may be understood better by considering meiosis.

First, with gametes and asexual zoospores inherited, meiosis occurred immediately after conjugation or, as Church wrote, at the first intent. Meiosis implies multiple fission. Therefore the zygote produced many zoospores but in course of time the number reduced to the minimal four. These haploid zoospores developed into haploid plants, which produced either asexual zoospores or gametes. This is the rather irregular life-cycle of many filamentous Chlorophyceae, both marine and freshwater, of which *Ulothrix* is an example.

Secondly, meiosis was delayed and a diploid many-celled plant growth intervened. Meiosis occurred at the formation of the zoospores or at the second intent, and the haploid zoospores grew into gametophytes. Thus may have come the regular isomorphic alternation of diploid, asexual sporophyte with haploid, sexual gametophyte, of which *Ulva* is an example.

Thirdly, by divisions of labour between sporophyte and gametophyte and extending thereby the exploitation of the environment, the heteromorphic alternation was established.

Fourthly, at any stage of evolution of the life-cycle, the evolution of sexuality to oogamy could occur.

Fifthly, zoospore evolution led to the passive tetraspores, and this seems generally to have followed oogamy.

Sixthly, the retention of the egg in the oogonium lead to the introduction of the third stage in the life-cycle known as the carposporophyte.

Meiosis occurs neither in the zygote nor in the next generation of carpo-spores, but in that of the tetraspores as meiosis at the third intent: two diploid asexual generations are inserted between successive haploid, sexual generations.

Lastly, an entirely different course could have been pursued without asexual zoospores. It leads to the diploid gametophyte of the fucoids, which, with meiosis at the second intent, produces haploid gametes. It parallels the life-cycle of animals and it seems to be the life-cycle, how-ever simple, of the diatoms.

Chapter 7

The Land Plant

IF THE STEADY, though buffeting, conditions of the shore have developed microscopic plants into swaying glades of seaweed, the land has improved this vegetation almost beyond recognition. By substituting, as it were, atmosphere for sea water the range of all the main factors of the environment that enable plants and therefore animals to live has been greatly extended. Life has moved from the hydrosphere, as the watery world is called, to the lithosphere of rocks in air. These rocks are not in constant wetness. Dried in the sun and cooled at night, their aridity cannot serve the foothold of plants. The sandy product of their erosion, which the sea washes away as shifting shingle, collects now on slopes in river flats as coarse or fine soil below which lies the water table and into which the rain soaks. The material of sterile beaches becomes the firm and fertile ground in which the majority of land plants prosper. Tender things, as rootlets, probe the porous soil and petals unfold. Epiphytes change from slimy tresses on flexing seaweeds to orchids on stately trees: vegetation is cast in a firmer mould.

Light-supply on land ranges from photosynthetic darkness, of course, to full sunshine, uninterrupted by sea water or intermittent tides. Temperatures range from $-70°$ to $77°C$, and both daily and seasonal fluctuations are much greater than in the sea. Water supply ranges from fresh to brackish, from ice to hot springs, from torrential downpour or perpetual mist to rare showers in deserts and drips in caves, but for most established plants it is the subterranean water table. Evaporation from the plant surface, significant only in the upper parts of the shore, becomes an over-riding problem. Mineral matter is available only to roots, but in the pristine content of all the rocks of the earth. Carbon dioxide and oxygen move with gaseous freedom. Winds mix and dry. The land itself is shaped into mountains, lowlands, valleys, ridges, plains, islands, and continents. Climates multiply [29].

How can one think of all the processes, physical and chemical, going on in the world that affect the lives of plants? A list would cover many pages and the active mind must be spared the tedium. Sunlight and water, then sunlight, water, and rock, and now, on land, sunlight, water, soil, and air make the three successive environments of plant life. If the four elements of the ancients are recalled, namely earth, fire, air, and water, here are ready clues to the scientific analysis of the plant's environment. Earth means foothold and mineral matter for plant growth, the variety of the solid state and rock chemistry. Fire embodies light and heat, which enter into all physics and chemistry, climatology, and therefore plant geography. Air signifies gases, their chemical nature, diffusion, evaporation, wind, and cloud. Water is the basis of protoplasm, comprising generally eight- to nine-tenths of its active and working state, and the manner in which water occurs determines the nature of plant existence. Finally, add strife, because living matter has inherited the capacity of the planktonic plant to behave individually and reproduce with that excessiveness that leads to mortal competition.

The wealth of vegetation on land is most difficult to reconcile with that in the sea. There are animals that make light of the distinction. Crabs, mudfish, turtles, penguins, and seals come out of the sea and in their limited ways enjoy the land, but such amphibiousness is denied to plants. The high tide is a barrier that neither seaweeds nor land plants may transgress. Seeds and spores sprout over the land, but no seed plant, fern, or moss, ventures down the shore under the waves; no seaweeds survive on land. The fresh water kills them, if not the exposure to strong light and desiccation: sea water kills land plants. There are some exceptions, such as a few microscopic fungi that grow on logs in the sea or as parasites on seaweeds, and there are flowering plants such as the sea grasses, which, by way of estuaries, have become adapted to marine life, not on rocks as seaweeds but on the detritus of mud and sand as land plants: but no plant commutes.

There is nevertheless a lesson to be learnt from zoology. Worms, molluscs, arthropods, and vertebrates from the sea remain worms, molluscs, arthropods, and vertebrates on land. They breathe air in place of water and may learn to drink. Arthropods have prospered into insects and spiders, fish into amphibians, reptiles, birds, and mammals, but no new sort of animal organization, no new phylum, has been evolved on land. The progress of animal organization has been in the sea. At various successive levels in this progress, animals have migrated to land so that a superficial idea of animal evolution may be obtained by comparing the several grades of organization in land animals; no zoologist, however, would suppose that the land vertebrates had been evolved from some minute beginning on land in parallel with the fish in the sea. With all its opportunities for living matter, the land has set such a hard and unconventional task for sea creatures that they have merely colonized it with the principles of construction laid down in the sea, and within their skins they keep its wetness.

Botany holds the same truth but it is concealed by the nature of plants, and it cannot be realized by looking in the sea for the seaweed relatives of moss or fern, such as fish is to frog, or lobster to grasshopper. These relatives are, by the nature of plants, extinct. The nature of the animal is to move, and to conquer and escape the environment as much as possible; therefore it may transform from sea to land. The nature of the plant is to become a static part of the environment and its interceptor. As there is no environment now between land and sea, there are no sea–land plants to intercept it. The ancient and extinct environment that brought plants to land had the plants which became the modern land plants; with the passing of that environment those ancestors disappeared. To look therefore for them in the sea, in the manner of zoology, is to misunderstand the whole botanical process.

A phylum, or main evolutionary sequence of plant life, begins with a planktonic state, goes through a benthic state, and then may colonize the land. It is defined photosynthetically and in the course of evolution acquires inevitably some or all of the common properties, both structural and reproductive, that characterize plants, be they green, brown, red, or even blue–green. Those plants that migrated to land did so with the sea heritage that they possessed and have remained at that sea level though it has become modified to land conditions. Thus we find repeated on the land plant structure and reproduction in simple, middling and advanced degrees and, as in zoology, this does not imply the evolution again on land of the characters evolved in the sea. A comparison will explain. From the types and customs of men in the Americas we may construct a progressive evolution of mankind from wigwam to UNO, but all are colonist from the Old World whence these things were brought.

Because of their green photosynthetic character, based on the predominance of chlorophyll, cane sugar, starch, and cellulose, green land plants belong to the phylum of green seaweeds. Some simple forms of these, such as single cells or filaments, have migrated on to land where, as microscopic green algae (which is a general name for all seaweed-like plants), they live in freshwater films on soil, rock, trunk, or leaf. In their simplicity they can be compared to ferns as worms to frogs. Ferns, however, in their massive parenchymatous structure with leaf, stem, root, and apical growth and with a hetero-morphic life-cycle such as in the brown parallel of *Laminaria*, repre-sent a much higher state of seaweed evolution that combines features of both laminarians, fucoids and *Dictyota*. Ferns then are to seed plants what amphibians are to mammals. The combination of sea characters which the fern ancestor implies are not commensurate with modern seaweed life, but they were, presumably, with the ex-tinct environment of the 'transmigration' to land. While we do not find, then, in the sea the ancestors of land plants, we do find in land plants the characters of their sea ancestors. Thus we learn that the best that the sea could make of plants is that which has fitted them best for land, again as the zoological parallel of fish into mammal. The less able either never succeeded on land or occupy what Church called 'inferior stations' outside the main progression of the better equipped.

It is important to realize this because the older books, written before Church had correlated marine and land botany, and many recent books that follow the old, subscribe to the view that there was a major evolution of green plants on land parallel to that in the sea. The great switch in understanding, which Church introduced, is the realization that land plants are not the outcome of such hypothetical and wishful thinking, but the land adoption, which can be under-stood, of marine plants that can be studied as photosynthetic evolu-tion co-operative with the sea. It means in effect that with a know-ledge of the fern, which starts with so great a heritage, there can be seen at once the major aspect of the life of land plants that is the forest: we do not have to learn all over again the beginnings of botany at second hand from ponds and bogs.

To see what the land has meant to plant life, consider its over-riding requirement, namely photosynthesis. The bright sunlight and the free movement of gaseous carbon dioxide and oxygen must have been as stimulating to primitive land plants as to the early animals emerging from the sea. Whereas seaweeds deposit cellulose walls round their cells from the excess of sugar that they make, the land plant overflows. There are still leaves that excrete sugar over their surface, but the more usual specialization of this excretion into nectaries has led to the animal-pollinated flowers, many of which drip with sugary nectar. More important for the plants, however, is the continued deposition of the excess of photosynthesis in the internal cell-walls, just as in the seaweed, but now it makes the timber for the tree with powers vastly exceeding those of the seaweed.

A larger plant will shade a smaller. On land, as in the sea, there is the upward struggle for light, and the bigger plants survive. There is, however, an immense difference between sea and land growth. The taller and larger the land plant becomes, the further are its leaves from the water supply in the soil. The greater, too, is its need for self-support. The seaweed grows into the water which supports it, but the land plant must become rigid under its increasing weight. Moreover, the larger the land plant, the more is it exposed to drying up. It must be waterproofed to prevent desiccation and this will, as effectively, prevent it from absorbing the rain that falls on it. Waterproofing is performed by the oxidation and stiffening of fatty substances in the outermost cell-walls, and it covers the surface of stem and leaf with a microscopic, almost impervious, and chemically inert pellicle called the cuticle: it replaces the mucilage of seaweeds. The cuticle interferes, however, with the entry of carbon dioxide for photosynthesis and of oxygen for respiration. It must be pierced with holes minute enough to admit air without waterlogging with rain. Yet, however small, they will at the same time allow water to evaporate through them from the inside. These holes are called stomata or stomates (figures 39, 40), and, as a microscopic means of communication with the exterior, they can be closed and opened. Some land plants have, indeed, so stoppered themselves, as the large cacti of American deserts, that they hardly communicate with the exterior; by a chemical dodge they carry on photosynthesis internally and, as for water, they may remain fresh, even uprooted and lying on their sides, for three or four years. They are freaks, of course, at the limit of terrestrial endurance, and do not represent the majority of plants, which continually move water through the stem from root to leaf replacing that lost by evaporation. This flow of water is called the transpiration-stream and it implies pipes inside the plant and roots to absorb: it becomes the overwhelming activity of the forest, which through plenteous evaporation engenders its own storms.

The root is comparable with the hold-fast of the seaweed but, instead of being a compact disc adherent to the rock, it must develop as much surface as possible to absorb water. It cannot fan out like a leaf in the soil but must take the form of intrusive and branching strands: there must be a branching root commensurate with the branching shoot. A superficial root system, like the branching hold-fast of the laminarians, will catch surface water but will not reach the sub-soil and will provide little purchase for holding the shoot. Such may have been the primitive root system under heavy rainfall but, if so, it has given place to underground roots providing better purchase and reaching the soil water, which is richer in mineral matter for plant growth. Underground roots in the dark cannot photosynthesize. They must be fed from the shoot, and there must be therefore a downward food path in the land plant, just as in the laminarian, but one that will not conflict with the upward passage of water. Inevitably the successful land plant, growing larger, must become a two-way construction. The green shoot spreads branches and leaves upwards in photosynthesis; the white (or not green) root branches downwards in the soil to absorb water and mineral salts, and to support the shoot. The more each grows, the further apart are their growing tips and the greater become the problems of transport, for neither part fends for itself. Instead of a mere hold-fast from which the thallus spreads and absorbs over its surface (in the manner of the seaweed) all that is required for growth, there is a dilemma with two horns, one into the ground, the other into the air. Transport between the two parts is internal and the shoot is waterproofed from its surroundings. Thus the land plant must grow internally. Plant form, which began as superficial outgrowth into the surroundings, becomes a strictly inherited method of intrusion. Botany used to recognize, if vaguely, these two kinds of plant, namely the thallophyte which grew and absorbed superficially, and the cormophyte, which is the root–shoot system with internal accommodation. They operate so differently, the one externalized, the other internalized, that they seem to widen the gap between the seaweed and the fern. If the cormophyte should return to sea life, as the sea grasses have done, it remains a cormophyte in construction and adapts this form to work as a seaweed: such is the effect of inheritance.

How the land plant has solved the dilemma successfully is shown by the seed of a forest tree [30]. It sends at once a strong root perpendicularly into the ground; then the shoot grows. How the

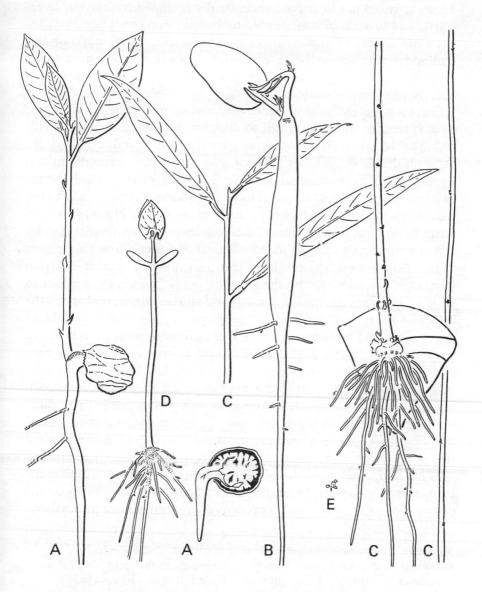

Figure 36. Dicotyledonous seedlings; × ½. A, Nutmeg tree *Myristica*, the germinating seed cut open to show the cotyledons in the food store (endosperm). B, Durian tree *Durio*, with one cotyledon removed. C, Mangrove seashore tree *Carapa granatum*, showing the seedling stem in three sections; numerous scale leaves below the three foliage leaves, and the terminal bud which is dormant at this stage for several weeks. D, Bean *Dolichos*. E, Minute seedling of a pitcher plant *Nepenthes*.

dilemma could not be solved successfully is shown by mosses, liverworts, and lichens, which, unable to form a strong root because they have not inherited sufficient sea progress, eke out a more or less thallophytic or seaweed-like existence on land by absorbing water over the surface when they can get it. These plants of necessity occupy the 'inferior stations'.

Growing up in botany means occupying more room. The first stem is slender and insufficient to support the branching shoot with its bigger leaves; it is not wide enough to carry the conducting system between the shoot and root. The problem has been overcome in two ways. Either, as in ferns and those flowering plants called monocotyledons, the stem apex enlarges and produces a thicker and thicker stem from the base of which new roots break out and thus, as in tree ferns and palms, a stout trunk bears big leaves and is maintained by copious small roots. Or, as in the conifers and those flowering plants called dicotyledons, the original stem must thicken after it has been formed in a slender way by the growing point. This is the manner in which most trees are made. A stout and thickening trunk carries the leaves on many small twigs and is supported by a few stout and thickening roots leading to the multitude of slender rootlets in the soil. In the fern and palm method the stout stem or trunk is a primary construction of the stout stem apex, and this method of growth is called primary thickening. That of the conifers and dicotyledons is secondary thickening by a special internal growing layer, called the cambium, and it leads to the formation of bark as the means of waterproofing the expanding trunk and main roots. Thus, in the upward struggle for light, trees have been made.

What happened beneath the heightening vegetation? Whether the primitive branches broke under their weight as they spread, or the plants died when they had reproduced or, continuing, petered out

Figure 37. Comparison of stem thickening in monocotyledons (primary thickening) and dicotyledons (secondary thickening), as shown by longitudinal sections of the stems; $\times \frac{1}{2}$. A, Apical bud of the full-grown coconut palm, forming inflorescences in the leaf axils, with the very massive apex protected by the leaf bases. B, Base of the trunk of a full-grown coconut palm, showing the numerous stout roots and vascular bundles (in black), and the base of the seedling superimposed in the centre. C, Diagram of the growth of a dicotyledonous forest tree, the cambium (*ca*) forming secondary wood (*sw*); the slender leptocaul bud, the thickening trunk, and the collar with the bases of the main roots, shown in four stages of thickening from the seedling (*s*).

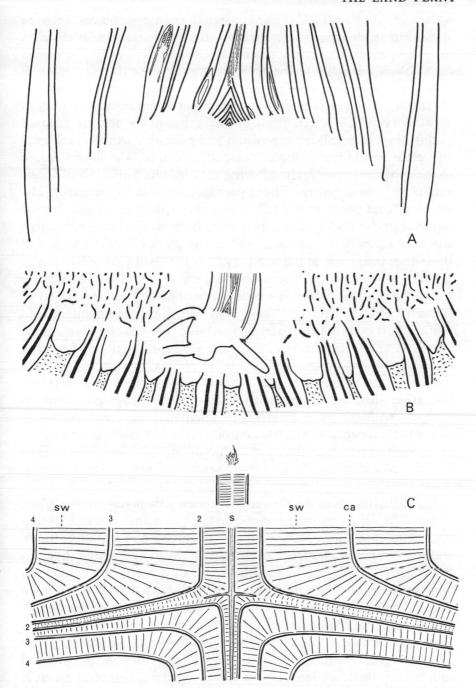

when they could not lift the soil water to the upper leaves, or were starved in shade by root competition, there was no tide, as cleanses the shore, to wash away the dead. Floods may have swept the pieces off to block periodically the rivers and start meanderings, even as now, but there must have begun to accumulate for the first time in the history of the earth under the vegetation an enormous quantity of vegetable refuse. Bacteria from the sea presumably became adapted to this new chemical decomposition and primitive worms, molluscs, and crabs would have nibbled, rasped, and torn; the floods would have deposited silt on to the growing rubble, which roots would have begun to bind together. Thus, perhaps, organic soil started. The spores of land plants would fall so that they sprouted not only on the top, but in the midst, of this primitive humus. A new environment was thus opened, not to the eaters or the photosynthesizers, but to those that could live in darkness by decay. It is the environment of the fungus, which is, as near as can be, a new sort of land organism. Yet it carries the marks of marine inheritance. It is composed of filaments of cells, without chloroplasts, which can be organized into tissues (particularly false parenchyma) like those of the filamentous seaweeds and it has marks of their marine reproduction. The fungi seem thus to have originated by loss of photosynthesis under those darkening conditions from the filamentous seaweeds of the 'transmigration' that, because of their feebler construction, were unable to compete with the more massive, parenchymatous, green plants. If not, what happened to the filamentous plants? Maybe an ingenious experimenter will, one day, put a chloroplast or chromoplast into a fungus cell and restore its photosynthetic existence: thoughts can be proved.

Howbeit, the fungus introduced a new character on the plant stage. It is called the saprophyte or the plant that lives on rotten (sapros) plants. Nowadays there is a triumvirate of bacteria, fungi, and invertebrate animals, all in great diversity, which live in, thoroughly exploit, and reduce to a fine tilth the fallen remains of land vegetation. When they began in geological time is not known but, to judge from the great accumulation of plant debris which makes the Coal Measures, either they were not then established or they were unable to cope with the chemistry of those plants. Their greatest development has accompanied the evolution of the final and much more putrescible forests of flowering trees. Primeval forest is not the orderly array of masts and spars over a clean floor that forest

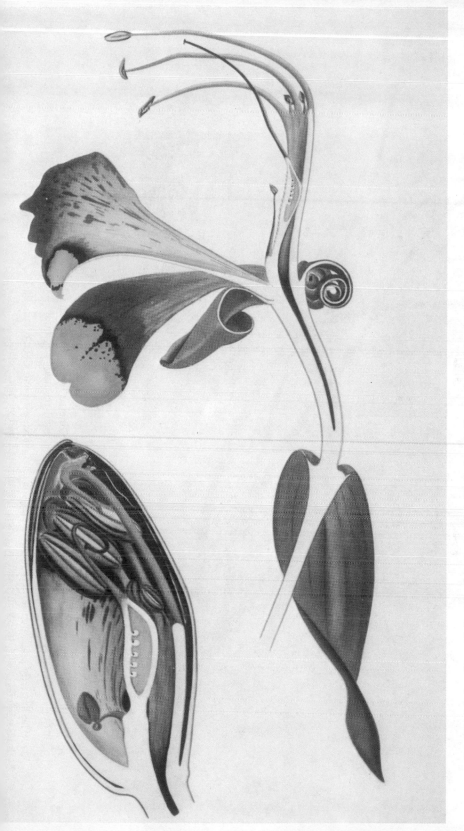

19. The bird-pollinated flower of the caesalpinioid *Amherstia* (Leguminosae) in section, nat. size; the bud, showing the construction, × 2.

20. The flower of the caesalpinioid *Intsia* (Leguminosae), pollinated by butterflies and, at night, by moths; one petal, ten stamens (seven sterile), nat. size: section, ×2; bud, ×3.

21. The butterfly-pollinated flowers of the caesalpinioid *Bauhinia flammifera* (Leguminosae), three stamens, ×1½ (top right) and ×3 (left); section and bud without petals,

22. Arillate fruits of *Aphanamixis* (mahogany family, Meliaceae);
twig, × ¹/₅; open and unopen capsules, × ²/₃; seeds with and without
aril, nat. size (lower centre); seed-section, × 2 (lower left); fruit
section, × ²/₃ (lower right).

23. Arillate fruits of the tropical lianes Connaraceae, ×⅔; seeds, nat.
size. *Roureopsis asplenifolia* (left); *Cnestis palala* (right); *Connarus
monocarpus* (bottom left, four figures).

24. Apocarpous fruits (fleshy follicles) of *Sterculia macrophylla*, × ⅓; seed-bases with vestigial aril, × 6 (top left and bottom right).

25. Arillate twin pods of *Tabernaemontana* (Apocynaceae), × ²/₃; seeds with aril, × 2.

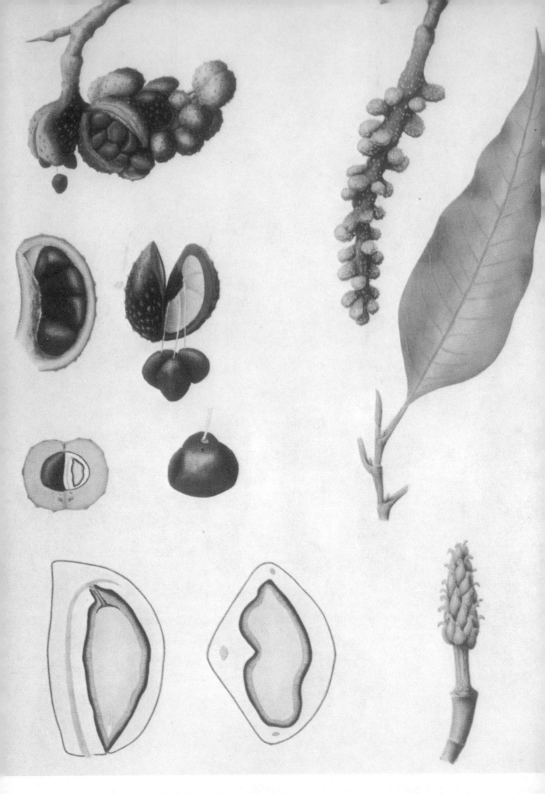

26. Apocarpous fruits (fleshy follicles) of *Michelia* (Magnoliaceae), immature (top right), mature (top left), and twig, $\times \frac{2}{3}$; three follicles, nat. size, and a seed, $\times 2$ (left centre); seed-sections, $\times 6$ (bottom left); flower-ovary, $\times 2$ (bottom right).

practice desires and to which we have become accustomed by commercialization. It is woodland where crowds of plants in a multitude of specific diversity vie in the endeavour to root and to grow up, and where large numbers are dying or dead, breaking up, crumbling, and falling as they are crowded out, pushed over, or pulled down by exuberant lianes. From the midst of green foliage project the dead limbs and skeletons of trees; on the floor is a tangle of rotting trunks, limbs, twigs, leaves, and fruits, among which saplings and seedlings are beginning. If the botanist would see the old drama again in modern dress, as on the shore, he must go to the tropical forests of flowering trees, where he will learn how nature in her intransigence has finally brought forth the most beautiful, the most artful, and the most destructive.

But, to return to the technicalities of land plants, ferns and seed plants have the most advanced reproductive methods of the sea, namely the fertilization of the egg *in situ* (attached to the parent, as in the red seaweeds), tetraspores (without flagella as in *Dictyota* and the red seaweeds) and an alternation of a diploid sporophyte, which is indeed the tree, with a small or microscopic haploid gametophyte (as in *Laminaria*). These methods, just as the vegetative structures, are now distributed among a variety of seaweeds, inasmuch as modern seaweeds have specialized livelihoods on the shore. The grand combination to fit the super-plant that migrated to land was not a requirement of the sea but of the historic passage from the primeval sea to the promising land. The rootless moss or liverwort adds to its humility a life-cycle in which the rootless sporophyte cannot escape from the gametophyte, which is the moss plant, but grows on it as a stalked capsule of spores reminiscent of the carposporophyte of red seaweeds. As we shall see, this is another failing which puts these plants out of the main progression on land: to use them as a prop in the theory of higher seed plants is like turning the earthworm upside down to make a vertebrate [31].

To recapitulate, the successful land plants, namely the ferns and the seed plants, have inherited the marine parenchymatous construction with apical growth, stem, leaf, and branching hold-fast, and a heteromorphic life-cycle with dominant sporophyte producing tetraspores and small gametophyte retaining the egg-cells in its tissue. In this light the new features required to fit the plant for land appear as details. They are: the waterproofing of the surface of the shoot with cuticle; the puncturing of the cuticle with stomata that

8

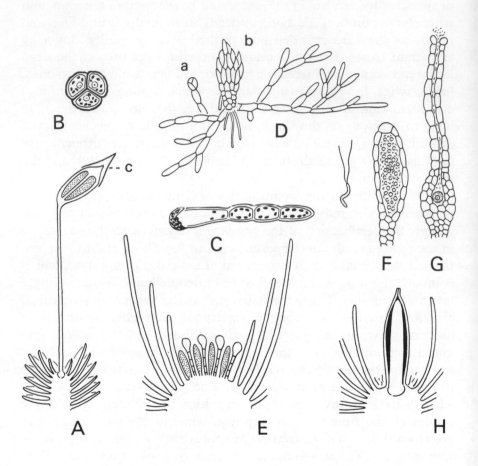

Figure 38. Incidents in the life of a moss. A, Section through the apex of a moss to show the stalk of the spore capsule (sporophyte) inserted into the apex of the moss stem (gametophyte); c, calyptra, or upper part of the enlarged archegonium, on the capsule; ×2. B, Group of tetraspores from the capsule; ×300. C, Spore producing a filament of cells with chloroplasts; ×300. D, Filamentous stage of the moss (protonema) forming a moss plant (b) and the inception of another (a); ×150. E, Section of the fertile apex of a moss to show the sex organs (dotted) among the sterile filaments; ×15. F, G, Sex organs, the male antheridium (F), and the female archegonium (G) with single egg-cell; ×90. H, Moss apex in section to show the young sporophyte developing from the zygote and still inside the enlarged archegonium; ×10.

lead to internal air passages; the internal position of the photo-synthetic cells, immediately below the surface layer; the rendering of the elongated internal cells into food and water pipes; the trans-formation of the branching hold-fast into roots fitted with caps to take the friction as the apex is thrust into the soil; the development of internal thickening and the internal positioning of the reproduc-tive cells, both in accord with the internalized growth, and therefore openings to let the reproductive cells out. These are the features of the flowering plant, introduced and learnt without explanation or understanding in introductory botany. To them must be added the procedure that has made the seed. The gametophyte, dependent on external water for the passage of the sperm to the egg, becomes more and more reduced until it virtually disappears in the flowering plant, and the tree, or sporophyte, forms on itself a new sporophyte, which is liberated in the seed. Thus the original double life-cycle, inherited from marine ancestry, is contracted on land into an apparently simple life-cycle, like that of animals, and the forest tree reproduces viviparously: maternal care is botanical, as well as zoological.

These are the chief structural alterations that turn the thallophyte of the sea into the successful cormophyte of the land. There are, however, four processes distinctive of the land plant; two relate to its manner of growth and two to its reproduction. The land plant perceives the weight of its own parts and its growth is controlled by the direction of the sunlight; it is sensitive, that is, to gravity and to light. Stems grow upwards and curve against the direction of gravity, roots downwards, while branches grow outwards and leaves are in-clined in various directions. Stems grow and curve towards light; leaf blades are set across the direction of light to intercept it, or have various oblique positions; roots may or may not grow away from the direction of light, though not towards it. There being no motion of nutritive sea water to sway the land plant, it is set more or less rigidly so that the shoot intercepts the light and the root enters the soil. The tree is a delicate balance of these processes but they are even more accurately employed in fungi, the fructifications of which, though lacking anything of the nature of an eye or of a pendulum, yet detect light and gravity with astonishing precision. Then, as the perfect touch to their composure, trees summon progressive animals with flower and fruit.

The engineering of the tree, the condensing of its inheritance into seed, flower, and fruit, and the devising of the fungus have been the main episodes in land vegetation. To perceive the accomplishment, let us view in marine retrospect what is, or used to be, a common sight in the tropical forest, for that is the botanical culmination.

There is a giant tree, pre-eminent in a forest that stretches to the skyline. On its canopy birds and butterflies sip nectar. On its branches orchids, aroids, and parasitic mistletoes offer flowers to other birds and insects. Among them ferns creep, lichens encrust, and centipedes and scorpions lurk. In the rubble that falls among the epiphytic roots and stems, ants build nests and even earthworms and snails find homes. There is a minute munching of caterpillars and the silent sucking of plant bugs. On any of these things, plant or animal, fungus may be growing. Through the branches spread spiders' webs. Frogs wait for insects, and a snake glides. There are nests of birds, bees, and wasps. Along a limb pass wary monkeys, a halting squirrel, or a bear in search of honey; the shadow of an eagle startles them. Through dead snags fungus and beetle have attacked the wood. There are fungus brackets nibbled round the edge and bored by other beetles. A woodpecker taps. In a hole a hornbill broods. Where the main branches diverge, a strangling fig finds grip, a bushy epiphyte has temporary root, and hidden sleeps a leopard. In deeper shade black termites have built earthy turrets and smothered the tips of a young creeper. Hanging from the limbs are cables of lianes which have hoisted themselves through the undergrowth and are suspended by their grapnels. On their swinging stems grows an epiphytic ginger whose red seeds a bird is pecking. Where rain trickles down the trunk filmy ferns, mosses, and slender green algae maintain their delicate lives. Round the base are fragments of bark and coils of old lianes, on which other ferns are growing. Between the buttress-roots a tortoise is eating toadstools. An elephant has rubbed the bark and, in its deepened footmarks, tadpoles, mosquito larvae, and threadworms swim. Pigs squeal and drum in search of fallen fruit, seeds, and truffles. In the humus and undersoil, insects, fungi, bacteria, and all sorts of 'animalculae' participate with the tree roots in decomposing everything that dies.

This tree is not alone. The forest consists not solely of its kind, nor even of a few, but of hundreds, each with its specific size, shape, bark, leaf, flower, fruit, and wood attractive to its particular following of green plants, fungi, and animals. Thus comes the richness of these forests, most prolific in natural history, and thence sprung the superiority of the creatures that intelligently traverse them.

The origin of these forests is not known. The geological record shows that they had modern form, except for specific trimming, at the beginning of the Cretaceous period. For over one hundred and twenty million years they have reigned, as long as or longer than any other kind of forest. Before them, from the Coal Measures onward, were forests of flowerless seed plants, now extinct, and of conifers and cycads, some of which survive. Earlier still were forests of horsetails, clubmosses, and ferns, relics of which also survive. Then, more than three hundred million years ago, the record of fossil forest dims [32].

Coniferous forests of pine, spruce, fir, larch, cedar, cypress, and juniper are 'the dark and gloomy forests' of Hiawatha, so unrewarding to animals and indeed to other plants. They stretch over a great part of the north temperate region, to the north of the broad-leafed forests of flowering plants, and at higher altitudes on mountains, to which they seem better adapted [33]. That they ever thrived in the tropics, before the flowering trees, is difficult to believe, yet there are tropical conifers, much less familiarly known as *Podocarpus, Dacrydium, Agathis,* and *Araucaria,* trees of which are scattered in the broad-leafed tropical forest. They extend into the south temperate regions, as the kauri pine (*Agathis*) of New Zealand, and the monkey puzzle and paranā pine (*Araucaria*) of South America. Both these genera, moreover, resemble the ancient Palaeozoic conifers. It seems possible therefore that living conifers are relics of a world-wide coniferous forest that preceded the era of flowers, and that, ever since, the broad-leafed forest has been ousting them. There is evidence that the struggle persists with the *Araucaria* forests of eastern New Guinea [34]; indeed from the Isle of Pines, off New Caledonia, which is a refuge of *Araucaria,* over the mountains of the Malay Archipelago and across Asia, Europe, and North America to Vancouver Island and Cape Flattery, where the Douglas firs stand, and the pockets of redwood (*Sequoia*) in California, there is, where man has not destroyed, the line of conflict.

To the older era of the seed ferns belong the cycads [35]. They do not form forests now but, in nine genera and about eighty species, they are also scattered in the tropical broad-leafed forest. They occur chiefly by the coast, on rugged hills, or in dry interior regions. Their fleshy seeds and large pinnate or fern-like leaves relate them with the extinct seed ferns that flourished roughly between 280 and 180 million years ago. The form of both is that of the tree fern. A stocky, unbranched or sparingly branched trunk terminates in a rosette of large leaves, which, together with the dead hanging leaves, cast deep shade. They are smothering trees, occupying much area for little stature: their water-piping and secondary thickening are inadequate for loftiness. Through such a low canopy broke the conifers with their small and needle leaves to exceed eventually by five or six times the seed fern or cycad forest. Nevertheless, so far as edibility went, the seed fern and cycad forest may have produced more animal food. Another relic of this period is the maiden-hair tree (*Ginkgo*) of China, the fatty seeds and kernels of which are gathered to this day and sold in Chinese shops all over the world; they must have taxed the jaws of Pekin-man (*Sinanthropus*), unless he broke them with a stone [36].

The conifers, however, were not the first lofty trees. About the time of seed ferns, yet surely of earlier origin because of their simpler structure and reproduction by spores, there grew tall forests, up to 120 ft. high, of plants that now are represented by the small club-mosses (*Lycopodium*, *Selaginella*), the quillwort (*Isoetes*), and the horsetails (*Equisetum*). They were composed of the giant clubmosses, as *Lepidodendron*, which had the aspect of *Araucaria* but retained in branching the primitive dichotomy of the seaweed and grew from immense forking root bases (the *Stigmaria* stumps of the Coal Measures), and the giant horsetails, as *Calamites*, which resembled bamboos in their long jointed internodes, evidently of rapid extension. It is difficult to see how these plants had anything to offer animal life excepting shelter for the hunted and the hunter.

The geological record does disclose a pageant of forest. It begins with practically inedible fern-like trees reproduced by spores. It leads through a long Palaeozoic phase of some one hundred million years during which the tree improved, seed evolved, and edibility increased. It culminates with the supremacy of the broad-leafed trees that offer flower, fruit, leaf, timber, and a general putrescibility for animals and fungi. Too much should not be read into the fossil record, however, because it is a record only of the tougher parts of plants. The soft parts would have rotted before fossilization, and this very putrescibility of early flowering plants and their predecessors may be the reason for their absence from the Palaeozoic rocks.

There is no clue, nevertheless, how plants came on to land. Without wood or cuticle, the bodies of the transmigrant plants were very unlikely to have been fossilized. A rocky, rainy, tropical shore overhung with cloud may be imagined, but this is too facile; there are such shores now, yet without a trace of the primitive sea–land plant that botany seeks. The problem is a great one and needs great thinking. Church considered that the facts of marine plant life pointed to a primeval ocean without land, the rising floor of which became colonized by the simplest forms of seaweeds and then, as this floor was lifted gradually into the air, so the bigger seaweeds were evolved and converted into land forms. The conclusion may be considered geologically unsound, but it is a botanical contribution to a very remote problem where little is proven, and botanists know that much of the history of the earth's surface is registered by the present occurrence of plant life [37]. There must have been a time and place when seaweeds grew into the air, such as they cannot do now.

But what use is this speculation ? Well, one day when the knowledge of the chemical organization of plants is advanced enough, they may be led backwards artificially, and botanical institutions will have, beside museums of fossils, conservatories of living antecedents. A new plant is a source of horticultural pleasure. Horticulture is a character of civilization, and the extinct plant revived will add to the pleasure. Without a consideration of the past of plant life, the new cannot be distinguished from the previous. Already there are signs that the qualities of the extinct are being bred into new generations.

Chapter 8

Tree-making

THE LEAF shows better than any other part of the tree the change on land that sea structure must undergo. Flattened and flexible stems are now out of order for self-supporting plants. It is the leaf blade that must retain the primitive shape and, with maximum ratio of surface to volume, compete for the effective absorption of light and carbon dioxide. As the most primitive and essential organ of the land plant it needs fullest administration. Land leaves differ enormously in size, shape, and manner of growth but there is remarkable uniformity in their microscopic structure. Once this had been settled, so it seems, then the leaf was able to evolve its various manifestations.

The surface of the leaf is completely turned into a transparent skin that firmly holds the internal structure and prevents rubbing and wearing by wind and rain. It consists of a layer of colourless cells, devoid of chloroplasts but invested by the waterproof cuticle. This epidermis, as it is called, is also a water-jacket, the internal photosynthetic tissue being in fact still surrounded by a water supply. In many plants that endure drought, as desert plants and epiphytes, the epidermis thickens into several layers of large water-storing cells; they form a tank or microscopic sea on which the photosynthetic cells draw. Water passes from the veins of the leaf to this epidermal jacket and so to the active cells. Just as the tenor of the ordinary human mind cannot be appreciated without the special revelations of the musical, for instance, the architectural, or the mathematical, so no ordinary part of a plant, such as a thin leaf, can be understood without comparison with the more select. Thus in contrast the underwater leaves of land plants that have taken to fresh water, such as water buttercups and water plantains, have at most an exceedingly thin, vestigial cuticle, not impervious to water and so, faced with an external supply, the epidermal cells are photosynthetic.

The tissue within the epidermis is called the mesophyll. It consists of three parts, namely the palisade layer, the spongy parenchyma, and the veins. Epidermal cells fit tightly without air spaces save at the stomata. The internal cells of the palisade layer and spongy parenchyma are largely separated by air spaces communicating with the

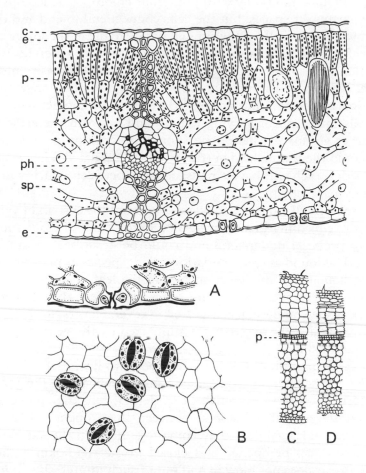

Figure 39. A small part of the leaf of the tree *Dillenia suffruticosa*, to show in section the internal structure with upper and lower epidermis *e*, strutted by the thick-walled fibres on the two sides of the vein; *c*, cuticle; *p*, palisade cells; *sp*, spongy parenchyma; *ph*, phloem of the vein (the xylem shown with thick black walls); stomata in the lower epidermis; a cell on the right with needle crystals; ×250. A, Stoma in section, and B, the stomata in surface view with guard-cells, chloroplasts, and air pore; ×500. C, Section of the fleshy leaf of *Peperomia* to show the water-storage tissue on either side of the palisade tissue *p*. D, *Peperomia* leaf after drying at room temperature for four days; ×20. (After Haberlandt, 1914.)

stomata. In the majority of leaves, the position of which is more or less horizontal, the palisade layer lies immediately below the upper epidermis and it consists of several rows of cells (one row in thin leaves) elongated at right angles to the epidermis, in the manner of railings, and lined internally with numerous chloroplasts. Round the cells are narrow air spaces, but the cells connect end to end and thus row to row with the cells of the spongy parenchyma in the lower half of the leaf. These cells, also photosynthetic though generally with fewer chloroplasts (corresponding with the lessened intensity of the light that has passed through the upper layer), are not elongated in any particular direction but are pulled out at their contact with adjacent cells into arms of varying length, and between the arms are much larger air spaces: hence the spongy description of this loose layer. Should the leaf be held vertically and illuminated fairly evenly on both sides, each will have a palisade layer between which will lie the spongy parenchyma.

In the mesophyll where its two layers join lies the system of veins. Patterns of veins differ greatly according to the kinds of plant and, if they were properly understood and could be described accurately, they would be found as characteristic of plant species as finger-prints of human beings. On the lower surface of the leaf, more often than the upper, the stomata occur. The edges of the leaf are toughened with thick-walled epidermal cells, often in several layers, to act as the hoop within which the leaf tissue is supported by the veins between the two tight skins.

A stoma is a mouth, 3–15μ long, between two lips, each of which is an oblong or semicircular cell called a guard-cell. When the guard-cells absorb water from the adjacent epidermal cells they bulge into them, and the air-slit, or stoma, between them opens: on losing water to the epidermal cells they come together and the slit closes. Thus stomata can be opened and closed in the rigid skin without buckling it. Air does not blow in and out of such minute slits, but its components diffuse through them according to their concentrations inside or outside of the leaf. In daylight, carbon dioxide is removed in the leaf by photosynthesis and more diffuses in through the open stomata from the outside. At the same time oxygen is produced by photosynthesis and it accordingly diffuses out. The concentration of water vapour in the wet interior of the leaf is nearly always greater than in the external atmosphere: therefore when stomata are open the inside of the leaf begins to dry and would dry up if it were not supplied by water from the veins. Vein-patterns are methods of leaf-irrigation.

A waterproof membrane stretched over a water surface will stop evaporation, but the effect of a membrane pierced with minute pores is unfamiliar. Provided that they are numerous enough and evenly spaced, the rest of the membrane has so little effect that evaporation

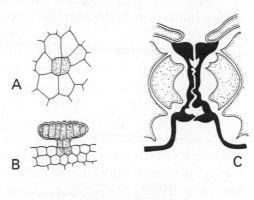

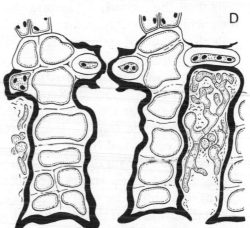

Figure 40. A, Epidermal cell of a leaf divided into two, which will become either the guard-cells of a stoma or the centre of a gland hair, such as B; ×400. C, Stoma of the nipa palm, with the narrow air passage further obstructed by cuticular outgrowths (the cuticle in black); ×800 (after Haberlandt, 1914). D, Stomata of a strangling fig (*Ficus glandifera*, New Guinea), deeply sunken in the lower epidermis (five cells thick) in a manner that retards evaporation; some of the pits leading to the stomata have germinating fungus spores; ×800.

is almost as great as from the free water surface. A given area of leaf, set with open stomata, evaporates therefore practically as much water as does an equal area of water under the same conditions. The leaf, however, controls very precisely the rate of evaporation by varying the size of the pores. Lacking muscles and soft tissues that can contract and extend, the rigid plant moves only by varying the size of its cells, and this depends on the extensibility of the cell-wall.

Generally, the thicker the wall the less extensible it is and, since walls thicken with age, so the young parts of plants are those that can change shape and direction.

Now this detailed structure, particularly of the stoma, suggests something entirely different from the marine. But, in biology, after examination of the adult structure, the principle is to study its development. In a young leaf while its parts are forming it is commonly impossible to distinguish the pair of cells that will become the guard-cells of a stoma from a pair which become a mucilage gland and excrete mucilage to lubricate the young leaf as it slips out of the bud. Mucilage excretion is, of course, a seaweed character and it seems that the guard-cells are modified mucilage glands of marine inheritance, which have done away with mucilage just as the cell-walls of the land plant are not mucilaginous. The air-slit between the guard-cells has the same nature as the air spaces between the cells in the interior of the leaf. These cells develop in a characteristic way. As the leaf begins to work, the inner cells towards the upper side divide into the palisade rows so that the uppermost row has more numerous cells than the second, the second than the third, and the innermost row, which joins the spongy parenchyma, has still more numerous cells than a row of spongy parenchyma, the cells of which are pulled out into arms to accommodate the expansion of the palisade layers. This is precisely the construction of the photosynthetic cortex of a parenchymatous seaweed. Replace the mucilage between the cells with air and cover the structure with a non-photosynthetic, protective skin, and the result is the microscopic structure, by no means recondite, of the leaf blade.

When the stomata open on a young leaf its internal cells are still enlarging and its air spaces expanding. It is possible therefore that during the unfolding of a leaf air is sucked into it through stomata. The effect is to turn the translucent green of the young leaf into the bright green of fresh foliage, in which much incident light is scattered by internal reflection from the cell-walls. Then, in course of time, the cell-walls thicken and become impregnated with waste substances from protoplasm, and with increasing opacity of the walls the foliage becomes dull and heavy. Leaves age and, after a time that varies from a few weeks in herbs to several years in palms, they die.

So far as fossil evidence reveals, this internal structure of the leaf has undergone only minor variations in the last three hundred million years. All tree growths possess stomata: they were requisite for internal aeration once the surface developed a cuticle, and plants without cuticle could not have emerged in active photosynthesis much above the water-saturated atmosphere at ground level. While the microscopic structure seems therefore to have been established very early in the origin of the land flora, there has been, by contrast, steady improvement in leaf-making. From the outset there appear to have been two extremes. The plant had either many simple leaves with a single blade each, which is the habit still of clubmosses and horsetails, or it had comparatively few large leaves each with many small blades, which is the habit of true ferns. The large leaves are called compound because they are made up of several blades, and pinnate because the midrib bears along it, like the feather of a bird, the blades (leaflets or pinnae, as they are variously called): if the midrib bears branches and these bear the blades, then the leaf is called doubly pinnate, and this process may go on to the fifth or sixth degree in highly compound leaves. Pinnate leaves show in their development how they have originated from a system of dichotomous branching with equal forks by the process of overtopping, such as turns the dichotomous fucoid into an axis, or stem, with leaves as in *Sargassum* (figure 24). Further, in living ferns it can be observed how these branches of the leaf become webbed together into larger units, and these units webbed again until finally all the leaf branches are incorporated in a single broad blade in which the midrib, side-veins, and network of smaller veins correspond with the original branchings of the leaf. A broad lamina is thus irrigated by continuous channels that preserve the original pattern, and from these patterns, so diverse in ferns and flowering plants, the way in which the leaf blade has been constructed and evolved may be learnt. The same process happens in the leaves of flowering plants, and there is an enormous and fascinating study, most inadequately explored, to be made of such problems as why a lettuce leaf, for instance, is quite different in detail from a tobacco leaf, and a gentian leaf resembles that of a plantain (*Plantago*).

The condensing of a branching system into one well-watered blade shows the way in which the extended, or protracted, and externalized growth of the seaweed has been contracted and internalized for the land plant. It is the chief way by which the peculiarities of the perfected flowering plants have come about; these peculiarities, commonplace as they may be in leaf, flower, and fruit, cannot be understood except through the 'explanation', the flattening and

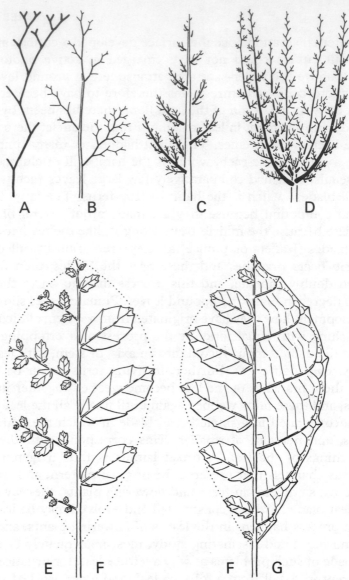

Figure 41. Diagram to show the derivation of leaves with limited growth, set in phyllotaxis on a stem, from a plant with dichotomous growth. A, Plant with equal dichotomy. B, Plant with unequal dichotomy and overtopping to form a main stem with short branches. C, As B, but with the branches (in black) arranged spirally round the stem. D, As C, but the main stem shortened (as in *Sargassum*) and developing long leafy branches resembling the compound (pinnate) leaves of ferns. E–G, Leaves of land plants showing the webbing of leaf branches into leaflets and the webbing of leaflets into a single lamina, the veins and teeth of which retain the manner of construction. E, Doubly pinnate. F, Simply, or once, pinnate. G, Simple, or undivided, blade.

extension, or the laying out in full detail of the corresponding structures of marine botany. Vein patterns are inherited structure adapted to the irrigation of the blade [38].

The stem is the cylindrical axis that produces and supports the leaves, and it bears the buds which become the branches. Usually, in seed plants the buds are in the axils of the leaf where it joins the stem and never on the leaf itself, which therefore never bears leafy branches: the stem can branch indefinitely, that is, but the leaf is a structure of limited size and shape. Leaf character is inherited and pre-formed in the bud, as fits the internalization of the land plant, unlike the extensive and thalloid fronds of most seaweeds.

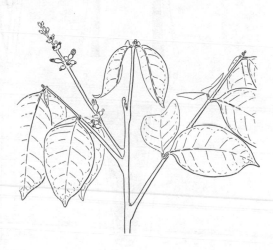

Figure 42. Tropical tree *Chisocheton spicatus* (Meliaceae) with pinnate leaves having at their ends buds that develop a pair of leaflets every time a new leaf forms at the stem-apex; × ½. This species develops eight pairs of leaflets over eight seasons before the whole leaf dies, but older leaflets are usually shed in pairs when four to five seasons old.

The cross-section of the stem of a dicotyledonous flowering plant shows much the same structure as the leaf, but the tissues are in circular patterns round the centre of the stem. There is a central pith, the cells of which may be pulled out into arms like the cells of the spongy parenchyma. Around the pith is a ring of veins that join those of the leaves at their insertions or at lower levels in the stem. Outside the veins is cortical tissue more or less separable into outer and more photosynthetic cells, and inner cells more as spongy parenchyma. The whole is bounded by the colourless epidermis with stomata.

The root, which is, of course, leafless, has much the same structure as the stem but it has no photosynthetic tissue or stomata, and usually no pith. The veins are combined to form a single central conducting strand, which, at the collar of the plant at ground level, opens out into the ring of veins in the stem. The absence of stomata may be connected with the prevailing absence of mucilage hairs and

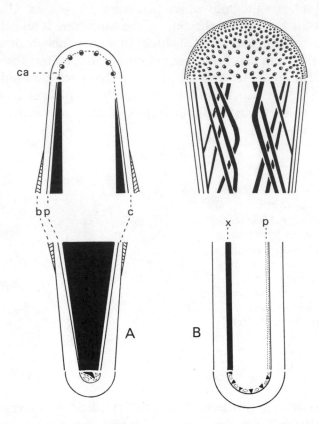

Figure 43. Diagrams of the structure of the stem (above) and root (below) of a dicotyledonous tree (A) with secondary thickening, and of a monocotyledon (B) with primary thickening; *b*, bark; *p*, phloem, shown by dotted lines; *c*, cortex; *x*, xylem, shown in black; *ca*, cambium, developing between the vascular bundles (shown by a broken line). In B, the primary thickening multiplies the number of vascular bundles. Structures are shown in transverse and longitudinal section.

glands on the hold-fasts of seaweeds; in other words, we should not expect them.

A longitudinal section of a root shows the growing end of the plant in the soil. The root-cap covers the delicate growing point behind which the cells organize themselves into the conducting strand and its surrounding cortical tissue. From the outermost cells, which have no cuticle, filaments grow into the soil and, as the root-hairs, attach themselves tightly to soil particles, just as filamentous hold-fasts may do, and they absorb the soil water with its dissolved mineral salts or, it may be said, such mineral salts as the plant requires for its growth. The effect is that from the many root-hairs, through the cortex of the root to its central core, then through the veins of stem and leaf, and finally to all the internal cells of the leaf there is a continuous wet system without air spaces through which water flows from the soil to replace that lost by evaporation through the stomata of leaves and

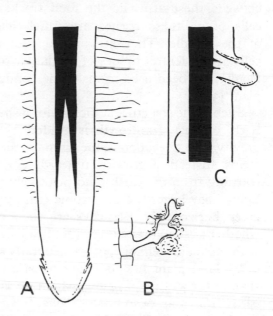

Figure 44. Longitudinal sections of a root to show: A, the root cap and root-hairs; B, the growth of the root-hair among soil particles; C, the internal origin of the root; the xylem of the root is in black. A, C, × 12; B, × 100.

stem. A simple analogy is the wick of an oil-lamp, but the plant is a growing wick, dividing above into new flames for which a branching base is needed to absorb more oil and a thickening strand is needed to convey the increasing amount of oil to the new flames. There the analogy stops, and the peculiarity of the plant continues uniquely. The stem must convey, also, down from leaf to root the sugars and protein constituents made in the leaf and needed by the root for its growth in the dark soil. The structure of the veins as two-way paths must be studied more carefully.

All veins consist of two, if not three, parts. There are dead cells that conduct the water: there are living, or half-alive, cells (without nucleus) that conduct the food substances; and there may be living or dead cells with thick walls that act as struts in the stem or leaf to sustain the other tissues. In the stem the water track, called the xylem, lies next to the pith, and the food track, called the phloem, lies next to the cortex; then, outside the phloem, or surrounding both tracks, occurs the mechanical strut or cylinder of thick-walled cells, known as fibres. The whole is generally called in botany the vascular bundle or the fibrovascular bundle. In the root the xylem of the vascular bundles becomes consolidated into a central fluted strand, which is star-shaped in cross-section, having two to seven arms, and between these arms lie the food tracks or strips of phloem. The cells of all these tissues are much lengthened in the direction of the bundle. They are indeed comparable with the elongated cells in the central tissues of the larger seaweeds, and this inherited faculty has been used to make the conducting channels of the land plant [39].

Such is the structure sufficient for a small land plant, as a seedling, herb, or fern. Its growth is guided by the reactions to light and gravity and they determine in part the form of the plant. Thus stems, instead of growing upwards and being negatively geotropic (turning from the earth) and positively phototropic (turning to the light), may become diageotropic (turning parallel to the earth) and creep horizontally. Branches are set more or less diageotropically whether as stem branches or root branches, and the angles they make with the main stem or root are generally specific to the plant. In fact, every land plant has its characteristic set, nowhere more striking than in the stature of palms, which, in general, are recognized more easily from afar, or in photographs, from the characteristic curves of the leaves than from the details of flower and fruit with which technical botany has aggravated their study [40].

Nevertheless there is still the question how the plant manages to maintain itself erect. Roots anchor it and the lateral roots particularly serve, not as props, but as stays. Yet the stem must have its own rigidity to support, whether in wind or calm, the branches and leaves. If a herb is unwatered it wilts; its leaves and stem soften and flag. Watered, it revives; the stem straightens and the leaves become erect from below upwards as the water ascends again to restore turgidity.

Every one of the thousands of cells composing the living tissues of stem, leaf, and root is a small bladder of protoplasm filled with water and surrounded by a cell-wall. The wall is elastic. As the protoplasm absorbs water it swells slightly and stretches the wall until it is taut, when no more water can be absorbed. Every internal cell thus presses on its neighbours as they press on it: in the spongy parenchyma the pressure is communicated along the arms of the cells. They press outwards against the epidermis where the cells are strapped together by the thick outer walls without air spaces. The tough skin holds the bulging interior, and the bulging interior supports the skin, just as a sack may be stuffed with balloons, the compression of which holds the sack rigid. As the air escapes from the balloons, the sack collapses: as the plant dries up, its cells shrink and it wilts. Thus in a very large degree stems and leaves maintain themselves erect merely through the turgid cells within the epidermis. Even the vast leaves of bananas and giant aroids are spread by this means; in the heat of the day when they lose more water than can be supplied by the roots they wilt and sag, but they recover as the sun goes down.

The fibres of the vascular bundles are another means of strengthening the green tissues. The fibres are much lengthened, needle-shaped cells with thick walls, tightly coherent without air spaces, set in the form of longitudinal strips. As the vascular bundles join at intervals along the stem, particularly at leaf attachments, the strips form a network or internal scaffolding of girders and struts that prevent twisting, bending, shearing, and shrinking, and sustain pulling. The fibres, obtainable from the plant simply by rotting the soft tissue, are the jute, sisal, hemp, rami, linen and other vegetable ingredients of string, cord, rope, or textile (excepting cotton). Once botany had satisfaction in the knowledge of these natural products of plants, but artificial substitutes and academic learning have removed their interest. Yet, the double-coconut, weighing fifty to sixty pounds, which is the fruit of the Seychelles palm *Lodoicea*, is suspended for six to seven years during the course of its ripening, chiefly by the fibres in its stalk. The trunk of the palm is sustained by these fibres. The 30 ft. leaf is mechanized by turgidity and fibres, and it resists the violence of tropical storms, which toss these massive structures about as if the crown would be blown off [41]. With eagerness, then, are the pages of economic botany turned to learn about the properties of fibres. At the base of many palm leaves there is a coarse brown cloth made of crossed and interlaced fibres clad in places with the dead epidermis; it surely provoked in the palm civilization of Asia the idea of weaving. It is the dead and dried system of fibrovascular bundles laid down in the leaf sheath at the base of its stalk, in a manner so intricate that it has never been explained. Indeed, but recently a Japanese botanist has explained how the vascular bundles are arranged in the comparatively simple stem

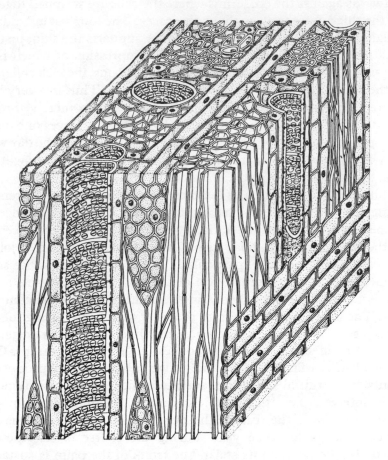

Figure 45. Block of wood of the Judas tree *Cercis*, showing the rays, wide vessels with close pitting, and the many narrow and elongate fibres; highly magnified. (By courtesy of A. Fahn, Hebrew University, Jerusalem.)

of the maize plant, which nearly every botanist has studied to learn in the beginning of his scholastic teaching the internal structure of a monocotyledon [42]. Be assured that there is scarcely a turn in the study of plant structure but understanding is lacking, and the inquisitive will never weary.

The fibres prevent also the squashing of the veins, particularly of the phloem, during the growth and swelling of the internal tissue and during bending of the adult stem or leaf. The long phloem cells have thin walls and delicate contents. Xylem, which struts the phloem internally, has its own strengthening. The walls of its lengthened cells become impregnated with an amorphous and complex chemical substance called lignin, which is the cause of lignification or woodiness [43]. Fibres may also become lignified and this is reflected in their commercial properties, for it makes them harder, less flexible, and inclined to be brittle, like splinters of wood. Lignification may extend also to the cells of the cortex of the stem, which is the tissue external to the vascular bundles, and even to those of the pith. Thus it makes the stone of the plum, the box of the hazel and brazil nut, and the hard shell of the coconut, which is one of the densest of plant constructions. Progressive lignification, like progressive calcification in the animal body, becomes an important matter in fruits, rendering them box-like and indehiscent so that the seeds cannot escape, and also hard and inedible. The grittiness of the pear is a form of diffuse sclerosis from tiny clusters of lignified cells.

The cells of the xylem vary from narrow fibres, which are generally alive, to wide, though lengthened cells, which are dead and full of water in transport along the vein. These cells lengthen in the young growing parts of leaf, stem, or root; then they deposit the lignin on the walls and at the same time their contents become very watery and begin to die. Even in young, soft plums or coconuts, the stony or woody layer is at first a very watery living layer before it becomes lignified and densely dead. Wateriness of moribund protoplasm and lignification seem, strangely, to go together and thus they make the water-conducting xylem. The lignin is not laid down as a continuous sheet inside the cell, but in a characteristic manner that reflects the state of growth of the part of the plant and the inherent nature of the plant. In young, growing parts, whether of stem, leaf, root, flower, or fruit, the first cells of the xylem to lignify in the vascular bundles do so by strands of lignin coiled spirally round the inside of the wall or by hoops or rings of lignin. They are called the spiral thickenings and the annular thickenings of the protoxylem (first xylem). As the parts of the plants are still growing, the protoxylem cell is also stretched, and its thickenings become pulled out, yet they still act as struts which maintain the dead cell as a tube of water while allowing it to lengthen. Later cells may become either long, narrow fibres or the larger water-conducting elements, which have more extensive lignification yet such as to leave on their wall thin areas that appear as round or oblong (oblique and transverse) windows, though they are called pits. With their heavy lignification these cells do not lengthen after formation.

Attention to this difference between protoxylem and later formed xylem (metaxylem) reveals an important detail. In the root, xylem-making starts at the ends of the arms of the xylem core, as seen in transverse section, and proceeds inwards. In the stem the xylem starts on the inner side of the vascular bundles, towards the pith, and proceeds outwards. In the root, the inward method means that the amount of xylem that can be formed is limited to the small space in the centre of the root: it limits the water-conducting capacity of the root and therefore the size of the shoot. In the stem, however, more xylem is formed at the sides of the bundles and the stem can accommodate more water-conducting tissue than the root. Now in some primitive ferns and fossil plants, in which these details of hard structure may be preserved with almost miraculous minuteness, the stem structure is the same as the root structure with inwardly forming xylem. It may have been the primitive state when stem and root agreed in structure, but it is unsatisfactory because of its limitation. Thus even if a leaf has only one vein and a vein only one pipeline of cells, ten new leaves mean ten new pipelines in stem and root, but both are already filled and conducting water at full capacity. The plant must stop growing or it must disconnect an old leaf every time it forms another. It will be a spindly thing unless it alters its internal construction, and this is what is found to have happened. Stem construction with separate vascular bundles placed towards the outside not only allows more xylem to be added laterally, but gives a better mechanical framework to the shoot.

In passing it may be mentioned that if some leaves, such as the dogwood (*Cornus*) or banana, are torn gently, fine white threads may be seen connecting the torn edges. With a pocket-lens they can be seen to consist of spiral threads drawn out, and they are the spiral strands of lignin from inside the cells of the xylem of the veins. Likewise the pink seeds of *Magnolia* dangle on white fleecy strands of lignin spirals torn from the minute veins that supplied the seed.

The vascular bundles of stems and leaves add more water pipes to their protoxylem by the division and multiplication of the cells which lie between the phloem and protoxylem. Thus rows of lignified cells make up the metaxylem and more can be added by the parenchymatous cells on the sides of the vein. As the leaf grows and makes more veins, which are collected into the midrib, the vascular bundles in the stem make more water pipes to connect with them. If the process

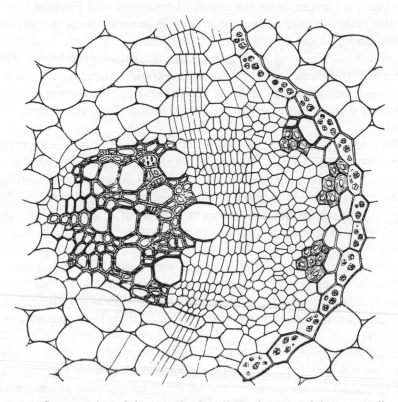

Figure 46. Cross-section of the vascular bundle in the stem of the castor-oil plant *Ricinus*, showing the development of the cambium from closely divided, small cells between the primary xylem (thick-walled) and phloem (thin-walled), the cambium beginning to extend across the cortical tissue to the neighbouring vascular bundles; highly magnified. (By courtesy of A. Fahn, Hebrew University, Jerusalem.)

continues all the vascular bundles will be joined laterally to form, in cross cut, a ring of xylem surrounded by a ring of phloem with a growing layer of cells between. If this growing layer persists with its activity, then it will form inwards a thicker and thicker cylinder of wood, which will supply the pipes to the increasing branches and leaves, and outwards the additional phloem for downward conduction to the enlarging root system. This layer, in fact, is that all-important structure called the cambium, on which the engineering of successful trees depends. It is the soft and sticky film where bark joins wood and indeed, because of its thin-walled nature, it allows the bark to be torn from the wood. The xylem and phloem formed by the cambium are called secondary in contrast with the primary xylem and phloem formed during the primary growth of the stem when it is constructing itself at its apex and forming the leaves. Since the primary xylem is an almost microscopic structure, the bulk of the wood formed in the stems of shrubs and trees is secondary and it is added during the years after the leaves on those parts have been shed.

Secondary thickening widens green stems into branches and trunk and it is continued downwards into the root. The cambium at the collar of the young plant, where stem and root join, is continued downwards between the strips of phloem and the xylem core of the root and round the flutings of this core to form again, as in the stem, xylem on the inside and phloem to the outside. Thus the branching shoot stands on a branching root connected by a thickening trunk which accommodates the increasing water and food requirements, and at the same time, by making a bulky core of wood from twig to rootlet, most massive in the trunk, the mechanical requirements of the growing tree are amply met: one tree may bear the weight of another.

The water pipes of the xylem are longitudinal rows of dead lignified cells, narrow in the first protoxylem, which supplies only the youngest tissues, and wider in the later-formed xylem, which meets the adult demand. Nevertheless, because of the cross-walls that separate one cell from the next, above and below, they are imperfect pipes. The cross-walls impede the flow. The final major improvement in tree-piping is the disappearance of these walls. They are dissolved chemically before the protoplasm of the cell is detroyed, and the remaining tubular parts of the cells become the sections of a continuous pipe. In some trees these pipes, which are called vessels to distinguish them from the long water-conducting cells known as tracheids, consist of few sections and are short. In others, many cells are involved, and the pipes range from an inch to several feet in length. In big climbers with slender stems, such as the rattans (climbing palms), the vessels may be 10–20 ft. long and carry water at considerable speed. Some vessels are narrow, others wide, and the widest occur, again, in climbing plants, where they may be 1 mm. in diameter. Vessels can usually be seen with the naked eye as the small holes in the cross-cut of wood. The tracheids, however, have microscopic width. Thus the wood of conifers, which has no vessels, is even and fine-grained compared with the more or less porous wood of broad-leafed flowering trees that possess vessels. Conifer wood, such as deal, has been found to offer three to four times more resistance to water flow than that of the broad-leafed tree, and this difference must handicap the conifer in competition

P   C   S   A   T   F   V

Figure 47. Typical cells of the conducting tissues of a flowering plant, the lignified walls shown in heavy black. P, Conducting cell (sieve-tube) of the phloem. C, Cambium cell with dividing nucleus. S, Tracheid with spiral thickening, and A, with annular thickening. T, Tracheid with the wall heavily thickened except for transverse pits, like windows. F, Fibre. V, Vessel made up of a row of heavily lignified cells with oval pits.

with the broad-leafed tree under a warm, wet climate conducive to rapid growth. Experiments have shown that water may travel at rates up to 70 cm. per minute in oak and ash trees, to 7 cm. per minute in beech and maple, to 0·5–2 cm. per minute in conifers, and at 1–2½ m. per minute in tropical climbers. These rates vary during the twenty-four hours according to the humidity of the air, the water supply in the soil, and the temperature; thus, under normal conditions, the rate will be zero just before sunrise when plant and air are saturated, and it will rise to the maximum between noon and 2 p.m. [44].

Secondary thickening occurred in the fossil tree clubmosses and horsetails, but it is absent from tree ferns, which consequently cannot develop a branching crown. Some modern ferns, even the bracken, have been found to have very simple and primitive vessels, but the great improvement of these structures into water pipes is a character of the flowering plants and must undoubtedly have added to their success as the major forest trees [45].

Secondary thickening is necessary to make a branching tree, but the wood that it makes has had to be improved. In the leaf it was the shape that had to be improved after the internal structure had been evolved. In the conducting tissues it was the minute structure that evolved after the general lay-out had been established. Here in endless details of wood anatomy, so interesting to botanists in every endeavour to relate the families of trees and to assess their abilities, and so important to the forester and carpenter in understanding the properties of timber, is one of the most enormous subjects of land botany. The student must turn to the work of Jane [46]. In this place there are but two other features of wood that need be mentioned. Firstly, dead wood is traversed by thin and usually microscopic strips of living cells that pass radially from inner to outer part and connect thereby the wood with the living cambium, phloem, and cortex. These strips are called the rays and, if the student wants to understand wood, he must dissect it under the binocular microscope. Secondly, in large trees the central wood loses its capacity to conduct water; its rays die but, before they do, additional substances are laid down in the woody walls that cause them to harden and often to darken. Thus the old central wood becomes the heartwood, and the young and pale peripheral sapwood conducts the water. The older the tree the thinner grows the shell of living wood, the greater is the content of heartwood. Here lies the source of death, for the heartwood is the part that fungi may attack after entry through the snags of dead branches.

The cambium is an internal layer of delicate non-photosynthetic cells, placed next the phloem (from which, presumably, it derives its food supply), and protected by the inner wall of wood. Nevertheless, it is remarkably sensitive to strains set up in the tree. How it detects them is not known, but it can react to compressions by forming more wood on the compressed side, as the underside of a branch, or it may react to strains with more wood as on the upperside. Branches sawn off near the trunk often show the pith eccentric, accordingly as more lower compression wood has been formed or, less frequently, more upper tension wood. At the base of the trunk, however, where it joins the lateral roots, tension wood is commonly developed into the thin flanges that have been called mistakenly buttresses. When the botanist–explorer Spruce first saw the buttresses of large trees in the Amazon forests he struck them, as many botanists must have done since, but only he observed that they gave out the note of a taut string [47]. By no means all trees can form buttresses, but in those that do the buttresses will be developed mainly on the side of most tension, for instance, up the bank by leaning riverside trees.

The outward growth of wood, which starts in the twig and converts it into the trunk or branch, stretches the cortex of the tree. In response to this strain the cells of the cortex lengthen sideways and begin to divide so as to maintain the surface unbroken. Generally cell-division is limited to a special layer of cells called the cork cambium. Wood cambium forms mostly wood on its inner side and comparatively little phloem on its outer side, for the volume of water conducted and evaporated far exceeds that of the food supply to the roots, but the cork cambium forms mostly bark tissue on its outer side and little or no living cortical tissue on its inner side. Bark cells soon die. Their walls become impregnated with corky material (suberin), similar to cuticle though not so waxy, and this severs them from living contact; their contents dry up and become filled with air. Thus bark acts as a dead waterproof covering and insulation to the thin layer of living tissue between it and the wood. If much dead bark is formed, it cracks and may scale off, especially if new and irregular layers of cork cambium are formed internally, as happens in many trees. The nature of bark varies greatly and opens another big chapter in tree anatomy, important to the botanist and the forester because in high and varied forest with many kinds of tree differences in bark supply the most ready means of identification [48].

In special places on the bark, which appear as paler spots, pustules or fissures, the tissue is loose, and even powdery. These places are the 'breathing pores', or lenticels, through which oxygen may diffuse for the respiration of the living interior. Slow indeed may be its entry but plant respiration is generally slow and slight, and the simple arrangement works. One way to kill a tree is, in fact, to suffocate it by blocking the lenticels; standing water or earth heaped round the trunk may be sufficient. In swampy riverside forest, however, the trees must be able to withstand periodic flooding: when the waters subside, there burst through the films of mud left on the trunks the new pale and powdery growths formed by the lenticels. Normally this powder of dead cells on the outside of bark is washed off by rain, but where it remains there may be a sufficient residuum of mineral matter in the cells to allow minute green algae to grow on the trunks. The smooth bark of the beech, continually sloughed in microscopic quantities, is thus often green with microscopic algae.

Bark cells may also contain substances that are obnoxious, distasteful, or poisonous to animals. Bark protects the trunk both mechanically and chemically from animal attack, whether from the larger biting animals or the small borers as beetles. When a tree is felled and its bark is smashed, there come hundreds of beetles, as if from nowhere, attracted by the released smell, to attack the wood. Similarly bark resists attack by fungi. That trees survive depends on the ability of the bark cambium to maintain unbroken the waterproof and protective covering of the thickening trunk.

It is sometimes said that a tree is killed by the removal of a girdle of bark. Cambium and sapwood are exposed and dry up, and this interferes with water conduction, while the cortex and phloem have been severed and the downward flow of food to the roots is interrupted. New bark can be formed by the outgrowth of new cambium at the edges of the girdle but the process is generally too slow. Nevertheless most of the many kinds of tree of tropical forest cannot be killed in this way. Into the cut a poison, such as sodium arsenate, must be poured to kill the living cells of the lower part of the trunk. Some of these trees have fibrous bark that can be stripped easily and it is used by native forest-dwellers to make clothing, baskets, and other utensils. The bark may be stripped all round the trunk for lengths of 10–30 ft., and yet that tree will regenerate new bark from the broken ends of the living rays of the wood. Such trees reveal the vivacity of trunks as well as our imperfect understanding of their organization. The botanist must refer to the tapping of rubber trees, which involves the skilful removal of bark external to the thin cambium, the regeneration of the bark on the tapping panel, and repeated tapping as frequently as the bark can be renewed [49]. The more that bark is studied, the more characteristic it appears of the species for it reflects the mechanics, the chemistry, and the inner working of the tree.

140

A tree can be summarized as a many-celled plant of block construction, the prolonged apical growth of which raises into the air through the perception of gravity and light a branching shoot, and drives into the ground by gravitational direction a branching root. The absorbing and feeding parts are the leaves and the rootlets, both of limited duration. They are renewed at the growing tips, and these tips, as they diverge, are connected by the older parts of stem and root. Absorption and lengthening have stopped in these older parts, but secondary thickening develops them into the support and conducting channel, protected by bark. The tree is organized by the direction of its growing tips, the lignification of the inner tissues, the upward flow of water in the lignified xylem, the downward passage of food in the phloem, and the continued growth of the cambium. It is kept alive, in spite of its increasing load of dead wood, by the activity of the skin of living cells.

Every one of these structures and activities should have a chapter to itself to do justice to its importance and to reveal the great amount of research and thought that have gone into its study [50]. The body of knowledge is vast, but the kingdom of plants is vaster. Understanding, in terms of physics and chemistry, relates so largely as yet to the particular species of temperate regions that it can hardly be welded into a general life of plants without overmuch assumption.

Chapter 9

The Upward Struggle

WILLOW BRANCHES spread widely. Narrow leaves are set outwards on short stalks, roughly in two rows along slender twigs. The sapling grows erect. Its leaves are borne in several rows round the stem, and the branches that come from buds in their axils grow upwards in all directions. As they branch in turn, the leaf arrangement becomes simpler and the twigs tend also to be set in two rows. Thus by branching to occupy space, and then by restricting the branching to consolidate the position, the tree builds a scaffolding that supports and exposes to the light with least overlapping the peripheral canopy of foliage. At length, long branches sag to the ground.

The poplar is the nearest ally of the willow. Its leaves are differently shaped. They hang and clatter on long stalks set round the stouter twig, which grows upward. The twig can branch in all directions but, with better illumination on the outside of the crown, the outer buds develop more strongly and raise the crown outwards and upwards. Inner branches, if they are formed, become overshadowed and in a year or two will die off. The erect growth of the twigs repeats the habit of the sapling. The lofty poplar therefore is more generalized in its tree-form than the willow. It keeps the all-round or radial construction of upward growth, whereas the willow changes to a flattened system of leaves and branches, making the sprays of foliage that are thrust across the path of light. Willows may weep but poplars express themselves with the stiffness of the Lombardy.

This distinction in growth and form occurs in many pairs of related trees. The beech resembles the willow but it is more emphatic in the lofty trunk, spreading limbs, and flat sprays of heavy foliage. Its ally, the oak, resembles the poplar, and the bends or 'knees' along its spreading branches are the joints caused by the successive outward growth of new twigs, the upward ends being overshadowed. So the birch is to the alder, the cherry to the plum or apple. In the tropics, the sapodilla or chiku (*Achras*) explains the poplar, and the tropical 'cherry' (*Muntingia*) the willow.

Cherry and plum belong to the same genus, *Prunus*. Birches (*Betula*) are so close to the alders (*Alnus*), and willows (*Salix*) to poplars (*Populus*), that in the vegetation of northern Asia and North America these pairs practically intergrade. Their distinction in tree habit must therefore have evolved not through divergent genera but within the limits of a single genus. As one proceeds towards the tropics, where tree vegetation becomes much richer in species and genera, the fact becomes obvious. Thus *Magnolia*, hollies (*Ilex*), maples (*Acer*), figs (*Ficus*), breadfruit trees (*Artocarpus*), oranges (*Citrus*), potato trees (*Solanum*), *Terminalia*, *Elaeocarpus*, *Cassia*, *Acacia*, and so on, have species of both kinds. There are fig trees like willows, and the holy pipal of India (*Ficus religiosa*) clatters like the poplar: there are beech-leafed figs and oak-leafed figs. Yet other tree alliances are wholly specialized in one way or the other. The nutmeg family (Myristicaceae), the custard-apple family (Annonaceae), and the ebony family (Ebenaceae), have the willow habit. The ivy family (Araliaceae) and *Bignonia* family (Bignoniaceae) have, like the rhododendrons, the poplar habit [51].

The distinction is the end of the long story told by all sorts of trees about their upward struggle to maintain a place in the rising canopy of the competitive forest. The end is the elegant willow pattern, but the start was awkward and humble. No great tree can have sprung into existence. Its engineering is too complex. Little by little low trees were improved to edge higher and higher than each other. Many of them failed and it happens that many of these survive, stuck at different levels of improvement, which have fitted into suitable niches in the forest. The botanist must live, however, in tropical forest to discover how modern trees have come about, what tree life really is, what vegetation can do, and, of course, what it has meant to the animals [52]. Long ascending limbs of poplar sort can be climbed, but with horizontal limbs animals swing, jump, glide, and perch. As schoolboys we can learn much about trees by climbing, and recapture their elevation. Then, for my own part, I was taught not from books, but by monkeys, flying lemurs, and other arboreal animals.

Orange leaves, small and simple, are jointed where the blade meets the stalk. The allied *Poncirus* has trifoliate leaves and each leaflet is

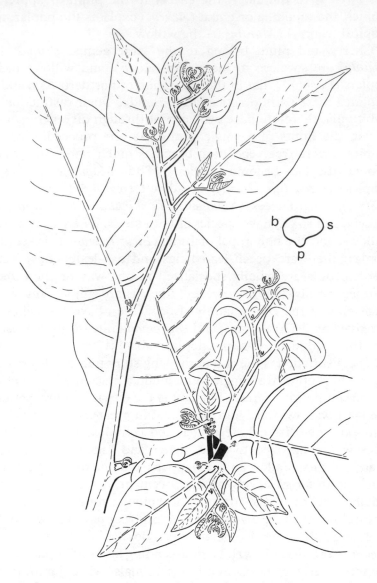

Figure 48. Common evergrowing tropical tree *Canangium* (Annonaceae) as seen from above, showing the leaves set spirally on the main stem (in black) and alternately in willow habit on the branches. On the main stem the branches grow at right angles to the direction of the leaves, from the axils of which they arise; two to three axillary buds, the uppermost becoming a branch; the branches show zig-zag development reminiscent of overtopping; $\times \frac{1}{2}$. Inset, section of the stem at a node; *b*, branch; *p*, leaf stalk; *s*, stem; natural size.

144

jointed to the top of the stalk. Other members of the *Citrus* family (Rutaceae) have pinnate leaves with several to many pairs of leaflets, each jointed to the main stalk. The joint marks the leaflet of a compound leaf. The simple leaf of *Citrus* is, in fact, no other than the end leaflet of the pinnate leaf when no other leaflets are formed: it is a

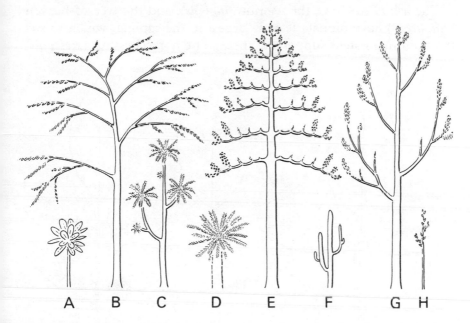

Figure 49. Shapes of trees. A, Cabbage or rosette tree, pachycaul with large simple leaves. B, Leptocaul tree with flattened sprays of leaves in two rows on the twigs. C, Branched rosette tree, pachycaul with pinnate leaves. D, Tree fern and cycad habit, pachycaul with pinnate leaves. E, Leptocaul tree with ascending twigs of spirally arranged leaves, building branches of regular outgrowth, as in oaks, sapodilla, or pines. F, Leafless rosette tree, cactus or spurge. G, Leptocaul tree with ascending limbs and foliage, as in the poplar. H, Sapling of G.

simplification. *Citrus* twigs are slender and incline to the willow habit. The pinnate leaves of the family are set spirally on stouter erect twigs with poplar effect. The *Citrus* family has the distinction in tree habit but the larger leaves of poplar habit are compound leaves. In the preceding chapter, it was indicated briefly that the simple leaf is the result of webbing of the compound, but the oranges show that it may also be a single leaflet of the compound: this leaflet, however, is already webbed from a more divided state. The point is that simplified leaves tend to be associated with the advanced willow habit, larger and compound leaves with the more generalized poplar habit. Do trees with compound leaves then have more primitive shapes ? There is no difficulty in finding the answer because there are many examples.

The ash (*Fraxinus*), the walnut (*Juglans*), and the tree-of-heaven (*Ailanthus*) have pinnate leaves, largest in the sapling, which shows the stout main stem supporting them. The horsechestnut (*Aesculus*)

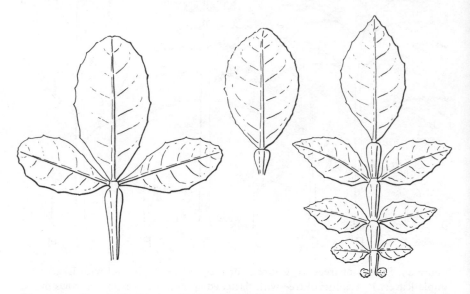

Figure 50. Orange leaf (centre) with the blade jointed to the winged stalk and representing the end-leaflet of a pinnate leaf (right, *Merrillia*), with the trifoliate leaf of *Poncirus* (left) as the intermediate; natural size.

has compound leaves, though they are palmate, which is merely the special state of the pinnate leaf in which the main stalk has not lengthened; and its large buds and stout twigs show the same massiveness of support. The bigger the leaf, particularly if it is compound, the greater is its demand for vascular bundles and the bigger must be the twig that accommodates it. Twig size and leaf size are intimately related, and as intimately connected with the habit of the tree. No plants show this better than the breadfruit tree (*Artocarpus*) and its allied species in the eastern tropics. The biggest pinnate leaves occur in the timber tree *Artocarpus anisophyllus*: they are spirally set on very stout ascending twigs. The breadfruit itself (*Artocarpus incisus*) has pinnately lobed leaves on twigs almost as big, whereas the species with small leaves have them set alternately along relatively slender horizontal or drooping twigs.

In this genus *Artocarpus* there is a transition from the primitive state with compound, spirally (or radially) arranged leaves on trees of poplar habit to the advanced state with simple alternate leaves on trees of willow habit, but the first state differs from the poplar itself because the crown is composed of fewer and much larger leaves on fewer and much stouter twigs. *Artocarpus* introduces the problem of the size and number of stems and leaves, and this leads to the considerations whether big leaves or small leaves make the crown a better interceptor of light, and which method is the more economical and efficient.

Obviously, little leaves can be fitted together into a closer mosaic without overlapping than big leaves. The tree with small leaves would gain in photosynthetic ability so long as it could maintain the increase in water supply that the greater area of foliage necessitates. The tree with flattened sprays of foliage, such as the beech, intercepts better and casts greater shade than the tree with ascending twigs, such as the oak. Further, little things can be replaced more cheaply than big things. Damage to the tree with little leaves and slender twigs is repaired more cheaply and quickly than in the tree that has to make massive replacements. Provided that the plumbing is adequate, the willow or beech habit comes out on top. Why then are there trees of other kinds? Now the meaning of tree evolution can be appreciated. Primitive trees started inadequate in engineering and in form because they evolved from land ancestors that were not trees but had inherited a certain manner of growth from sea ancestors. How they could have turned this into the tree can be learnt still in the profusion of the tropics, though it happened in the main so long ago that no one can guess to within a hundred million years when trees began.

As another example of these principles, the bean family (Leguminosae) may be mentioned. It is distinguished by its pinnate foliage, but it has modified this foliage also into the willow habit while retaining its compound character. It has trees with large, twice-pinnate leaves, spirally arranged, with large leaflets (pinnae), difficult to fit into a close mosaic: the crown is rather open and ragged. These leaflets can be reduced in size and fitted more compactly, even with almost angular outline in geometrical precision, and the whole leaf can be diminished so that it can be borne in alternate arrangement on a slender twig. Thus arise the compact, regular, but well-ventilated crowns of *Acacia*, *Cassia*, and their allies.

Botany has to reckon, however, with leaves still bigger and more compound than those which have been mentioned. There are doubly pinnate, trebly pinnate, and even four-times pinnate leaves,

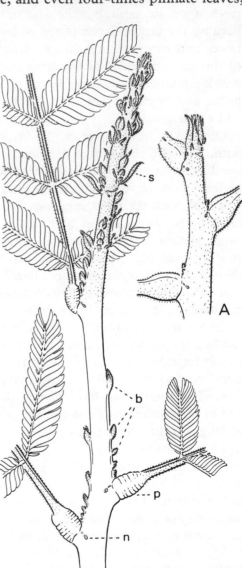

Figure 51. Elaborate twig of the tropical rain-forest tree *Pentaclethra filamentosa* Leguminosae, America); *n*, nectary at the base of the leaf; *b*, numerous axillary buds, which may become branches or inflorescences; *s*, scale leaves, each with two stipules, borne on the main shoot at the beginning of its new growth before the doubly pinnate foliage leaves; *p*, swollen leaf base (pulvinus), which adjusts the position of the whole leaf; natural size. A, Resting bud of the African *P. macrophylla*; natural size.

such as are familiar in ferns and the fern-like leaves of the carrot family (Umbelliferae). But these large and more compound leaves occur also on trees fitted, of course, on to stouter twigs with ascending growth. The student of tropical trees will begin to notice that among the ivy trees (Araliaceae), the *Bignonia* trees (Bignoniaceae), the meliaceous trees, and even the leguminous trees with such big leaves, the tree itself is actually of lesser stature, has fewer branches, and begins to look imperfect. It seems a general rule that such very elaborate leaves are too difficult or too gross to be engineered into loftiness. They can be borne on trees up to 50 ft. high, even 100 ft. high, but not 150–250 ft. high, which is the run of the canopy of lowland tropical rain-forest.

Proof of this difficulty is given by the saplings of some lofty trees. While still young, up to heights of 12–30 ft., they bear large, even very large, compound, often highly compound, leaves and then quite suddenly switch over to smaller simple leaves: they do the trick that the others cannot. A breadfruit ally (*Artocarpus elasticus*), which grows as big as any species of its genus, has sapling leaves up to 6 ft. long and four to five times pinnately lobed; when 20–30 ft. high the main stem and branches switch over with brief transition through webbing and shortening of the lobes into one smaller lamina. In other cases (Sterculiaceae), large palmate leaves with many leaflets pass into adult foliage of simple leaves representing the single median leaflet, as in *Citrus* trees, or a completely webbed leaf. There is some evidence, too, that at this critical height root-pressure, as a means of elevating water to the leaves, becomes less effective and even fails.

Comparison of these trees in other respects with their immediate allies confirms the theory that the massive twig with large, compound leaves turns into the many small twigs with small and simple leaves. And, as would be expected of the more primitive, trees with such massive construction are far fewer in kind and number in wild, untampered forest and occupy less eminent positions than do the advanced with poplar and willow habits. They are in their way relics of a stage in the evolution of broad-leafed flowering forest.

The history of the willow pattern leads further back. There are yet simpler trees, 10–30 ft. high, unbranched or with a few branches, which grow straight up and merely repeat the character of the main stem. It peters out at this low height; the leaves grow smaller and dwindle to incompetence; no crown is constructed. The trunk is soft; lignification is so slight that the trunk can be cut with a pen-knife; secondary thickening is weak. As trees they are so feeble that they are often regarded as overgrown herbs. They have been dubbed umbrella trees and rosette trees from the umbrella-like rosette of large leaves. Their timber is useless; their appearance is unornamental; their parts are too large for the herbarium; they are intractable and generally unwanted; they are very little known and discarded as curiosities. They occur, usually with rarity, among diverse families of plants where the student of trees can recognize in them the same primitive rarity as the monotreme in zoology or the cycad in botany. One kind nevertheless is familiar as the American pawpaw or papaya (*Carica*) now widely cultivated in the tropics for its melon fruit. There are other species of *Carica* that are better trees, up to 50 ft. high, more branched and, as one has now learnt to expect, with smaller and even simple leaves.

We have now the general range of tree form in modern broad-leafed flowering forest. There is the short, unbranched, or sparingly branched tree with large branching leaves borne on a stout stem that thickens relatively little. With the habit of a tree fern or cycad, it is a smothering plant, based on the principle of the large, primitive form of the leaf, and it gradually raises its comparatively small but dense crown of leaves from ground level until it can be engineered no higher. The other extreme is the lofty, canopy tree, the immense crown of which consists of thousands of small leaves borne on hundreds of small twigs, a few of which thicken into limbs while most are discarded; it is supported on a stem which secondary thickening converts into a pillar. It starts as a lanky sapling that fairly quickly overtops the smaller growths of the forest. Between the extremes survive all manner of intermediates variously improved in branching, leafing, heightening, plumbing, thickening, lignification, bark development, and other points necessary to complete the big tree. They survive because the taller have as it were lifted the forest canopy and left beneath the scaffolding of trunks and limbs all manner of gaps that the intermediates fill in. Some thrust branches with radial construction up the shafts of light that penetrate between the large crowns; others, such as the nutmegs which abound in tropical undergrowth, thrust more elegant willow sprays across the shafts of light.

The root system, also, must not be forgotten, though little is known about it because of its underground nature and the difficulty of disentangling forest roots in such profusion. The tree cannot improve without bettering its root system. There is evidence that rootlets have also become thinner and more branched from a stouter and less branched primitive state, which is shown, for instance, by the screw pines (*Pandanus*) and the palms in contrast with the thread-like rootlets of rhododendrons.

The beginning of broad-leafed forest must have been very different from the finality in which it seems to have persisted from the onset of the Cretaceous period. An undergrowth of smothering rosette trees may be imagined in gloomy coniferous forest, or such vegetation may have developed as dwarf forest in the open where new soil was forming by riverbanks or on landslips. The original rosette trees were either shade-growths, as some now are, or sun-growths in the open where most of such forms may still be found in various improvement, as the common large-leafed saplings and moderate-sized trees of such as *Cecropia*, *Ficus*, *Macaranga*, or *Hibiscus*.

From plates 18, 27–29 the meaning of the upward struggle to the limit of tree engineering can be gathered. Plate 18 is the papaya rosette tree, which a beast could browse on, knock over, and chew up. Plate 27 is a lanky pinnate-leafed tree with stout, sparingly branched twigs and little spread of crown; an animal must climb it and come down for there is little chance of passing from crown to crown. Plate 28 shows the taller tree with large but simple leaves, thinner and more branched twigs and more spreading limbs. Plate 29 shows the lofty dipterocarp tree, which represents the culmination of tropical rain-forest in Malaysia, where many species of the family Dipterocarpaceae construct the most luxuriant forest on earth. The immense trunk, superbly engineered and, consequently for its destruction, in high demand for timber, branches at a height of 100 ft. into the large canopy of small leaves, many twigs, and spreading limbs along which animals may travel from tree to tree, and eat, sleep, and give birth without returning to the ground. That is tree life. The small leaves glitter, reflecting light and transmitting sun flecks into the depths beneath, where the slanting rays of the ascending and descending sun penetrate. Plate 39b shows the airiness of the great crown.

Where tropical forest has been cleared by hand, for the bulldozer smashes everything, a few big trees are left in the open because they are the ironwood trees from which the axe rebounds. In picturesque grandeur they relieve the sad monotony of human endeavour until by decay, against which their hard timber is no guarantee, they too succumb. We used to call them tropical dendrons, so gracefully pruned in forest shade of the lower limbs, and saw in them glimpses that the older artists caught of the passing primeval forest of Europe. Bereft of the close surround, exposed to wind, and serving for the roosts of the birds that bring the seeds of destructive mistletoe and strangling fig, they die back, hollow, and decay. To catch these dendrons in their prime is a duty of the botanist; they may not be seen much longer.

Flowering plants are classified into natural families that are distinguished mainly by differences and resemblances in the structure of flower, fruit, and seed [53]. They are called natural, as opposed to artificial, because it is these resemblances which the theory of evolution explains as the natural outcome of descent with variation from a common ancestor; every family therefore, if properly classified, should have its common ancestor of the flowering plants. Large families such as Leguminosae, Moraceae (mulberries, breadfruit, figs), Anacardiaceae (mango trees, poison ivy), Apocynaceae (oleander, periwinkle), and Euphorbiaceae (rubber tree, castor-oil, spurge), possess all manner of tree forms. Tree evolution must therefore have occurred independently in them all. The family characters must have been evolved by the ancestors when they were lowly rosette trees. Then through the upward struggle they contributed their own kinds of tree to the forest. A leguminous tree for instance is the parallel, not the descendant or ancestor, of a moraceous tree, as both are parallel to the mango tree or the giant chewing-gum tree (*Dyera*, Apocynaceae). Then, in most large families, lianes, creepers, shrubs, and herbs have also arisen in parallel and become adapted to temperate regions, mountains, deserts, and so forth. The botanist is driven for clarity of research to think of plant life on the special lines of classification. There is no short cut from a buttercup to a poppy but one must think back to the ancestral plant form of their families and discover their parallel, and maybe complementary, modifications into the herbaceous state. With the parallelism of plant evolution we are already familiar from marine botany: it carries on with greater diversification in the land flora.

Where do the primitive trees survive? They can be found as already mentioned in forest shade, where they resemble the large-leafed saplings of more successful trees, and in open ground where they are the pioneers in full sunshine, among which the seedlings of the forest trees grow up to replace them. So much forest has been cleared now, and there are so many man-made waste places that these simple trees, originally rare in the ubiquitous primeval forest, have become some of the commoner. Others survive as epiphytes on the branches of the big trees and may send their roots down the long trunks even to the ground: it is the habit of several rosette trees of the ivy family (Araliaceae). Epiphytes live under exposed conditions of periodic drought more drastic than those of open places, which, in full sun and hot soil, are drastic enough compared with the humid shade of the forest. There is a tendency for these trees to have seeded and pioneered their way out of the forest into drier places.

Deserts have many spectacular umbrella trees, but they may not be easy to recognize at first sight, because, befitting the dry and often waterless conditions, the leaves particularly are modified. The cacti of American deserts and the spurge trees (*Euphorbia*) of the African deserts are exactly such umbrella trees, but they have lost their leaves and, by means of their thick tissue with poor lignification, their stems can become water stores. Similar odd trees occur in the Apocynaceae (*Pachypodium*, *Adenia*), and the vine family (Vitaceae, as African species of *Cissus*). It begins to be clear that many if not all of the odd-shaped plants of outlandish places are diversifications of this early umbrella kind of tree. Deserts are well known to harbour primitive plants that have become able to survive in their exacting conditions, out of the reach of the even more exacting struggle for existence in the dense forests. Such a survivor of very ancient vegetation, representing possibly something of the ancestry of flowering plants, is the strange, thick-stemmed, unbranched, and two-leafed *Welwitschia* of south-west Africa.

From deserts the botanists turns to other refuges, tropical mountains on the slopes above the tree limit but where frost is not severe, for thick-stemmed plants are too fleshy to be frost-hardy. Here are the strange umbrella trees, or tuft trees, of the Compositae, such as the tree *Senecio* of the African mountains, *Espeletia* of the Andes, and, in a dwarf form, the spearbush (*Argyroxyphium*) of Hawaii. Then, too, there are islands that have been long isolated, serving as backwaters from the main stream of continental evolution. For instance, on Juan Fernandes, St Helena, the Canary Islands, Socotra, and the Hawaiian Islands there survive many odd rosette trees or treelets of such genera as the sowthistles *Sonchus*, the houseleeks *Sempervivum*, and the sea lavender *Statice* (Canary Islands), the tree

153

gourd *Dendrosicyos* and the mulberry-ally *Dorstenia* (Socotra), and the tree geranium *Geranium arborescens* and the tree lobelias of Hawaii. Because they are such odd plants to those unacquainted with tropical forest and tree evolution, described in the words of Hillebrand for the tree lobelia *Brighamia* as 'a big cabbage-head stuck on a naked pole' [54], botany usually treats them as special adaptations. We assume that cabbages have been produced anew by cultivation but the cabbage character is just what the tropical botanist would expect for the progenitor of its family Cruciferae; cultivation may just as well have selected the characters reversionary to the

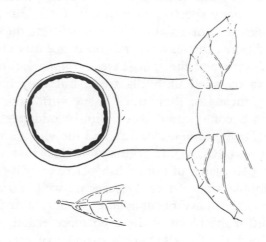

Figure 52. Difference between pachycauly and leptocauly as shown by sections of the young twigs of figs. Above, *Ficus salomonensis* (Bougainville Island), a pachycaul cabbage tree with wide pith (the xylem in black), the leaves six feet long. Below, *F. leptocalama* (North Borneo), with slender twig and small leaf of willow habit; natural size.

primitive, fleshy, and more edible state. They are special adaptations, but of primitive tree forms in their families surviving beyond the forest competition, which has far outstripped them. We need in fact to re-examine plant geography from the point of view of these plants [37].

To summarize the essence of forest evolution by flowering plants, there are two words to denote the beginning and the end. 'Pachycaul' (with thick primary stem) denotes massive construction as of the rosette tree or cabbage tree. 'Leptocaul' (with thin primary stem) denotes the slender willow construction. The pachycaul plants establish themselves by robust growth but they are expensive to maintain. Damage to the bud or leaf necessitates the construction of another large bud or leaf, as slow as it is massive. The leptocaul plant establishes itself by slender overgrowth of others and its small parts are easily replaced. It can survive with its numerous small dormant buds many set-backs that would be disastrous to the pachycaul. Big leaves and big sappy twigs are easily killed by frost. Small leaves may be shed in winter or the dry period of a seasonal climate. Small leaves can be developed quickly in spring; big leaves may be too tardy to catch the growing months. Thus leptocaul plants predominate in temperate and subtropical climates. They predominate in the tropics because they are the advanced state that has mainly transformed, or rebuilt, the major environment of plant and animal evolution since the mid-Mesozoic or, perhaps, the Palaeozoic era. These geological words, coined for events in the carnivorous past, have really little implication either for plants or for the vegetarian animals.

It will soon be three hundred years since John Ray, at the end of the seventeenth century, contributed a stroke of genius that took botany a century to appreciate. He realized that there were two radically different sorts of flowering plant, which he called monocotyledon (one seed-leaf) and dicotyledon (two seed-leaves). He inferred the distinction from the very limited number of seedlings that he grew. Modern botany has proved, maintained, and amplified the discovery [55]. It has added differences in leaf, flower, and internal structure, though none by itself is as distinctive as the number of seed-leaves. We have considered hitherto the dicotyledonous tree. Monocotyledons have several parallel tree forms, variously incompetent and therefore odd, so that they contribute little to forestry, and agriculture is their field. Nevertheless, the tree monocotyledons help greatly the understanding of the primary nature of the forest tree compared with the herb [56].

With a few exceptions, the monocotyledon has no secondary thickening. The stem contains many vascular bundles distributed throughout its tissue, though generally crowded near the surface and least crowded in the centre: there is rarely a pith without vascular bundles. It follows that the monocotyledon can form no trunk of timber and no stout tap-root. Its stoutness depends on the size of the stem-apex and root-apex, where in primary growth the parts are

formed. The monocotyledon is therefore either thoroughly pachycaul and massive, or it is variously leptocaul and slender. In either
case it must form new roots from the stem because the first small root
of the seedling has a limited amount of xylem and phloem that cannot be increased. Then, the leaf of the monocotyledon grows at the
base. Fern leaves, cycad leaves, and the more primitive compound
leaves of dicotyledons grow at the apex in the usual marine manner,
but some of them, especially the dicotyledonous, have been converted into basal growth. The leaves of cabbages, lettuces, and
thistles grow at the base, otherwise they would not be so suitable for
men and donkeys. Basal growth inflicts a parallel arrangement of the
main veins, but this occurs also in plantains (*Plantago*) and such
Compositae as *Tragopogon* and *Scorzonera*, which are dicotyledons.

Monocotyledons are usually considered from the temperate angle
of lilies, tulips, onions, iris, rushes, sedges, and grasses. These are
specialized leptocaul plants retaining remnants of pachycaul construction in their underground stems, though the textbooks regard
these parts as special storage organs adapted to the seasonal climate.
Far larger parts, replete with far more storage, make the normal
construction of the forest monocotyledons that in continuous growth
maintain the primitive and massive smothering habit. For example,
there are the palms, the screw pines (*Pandanus*), bananas and their
allies, and the large aroids, as well as sub-tropical plants as the Nile
sedge *Cyperus papyrus* and the Australian blackboys *Xanthorrhoea*.
They may seem oddities, but in their light the study of monocotyledons must be entirely recast.

Palms are umbrella trees, rosette trees, cabbage trees, or pachycaul
trees *par excellence*. They have succeeded in the forests by perfecting
the habit with enormous leaves 1–30 ft. long, which take six months
to five years or more in which to grow. Thus no palm would become
deciduous and very few, like the Mediterranean *Chamaerops*, the
Japanese *Trachycarpus*, and the Andean *Ceroxylon*, can stand frost.
All palm seedlings start with an onion-like structure on which they
erect the massive primary stem, and the best onion appears as a very
diminutive, or neotenic, palm limited to one or two season's growth
in the open. How forest herbs have been derived from the pachycaulous state is shown even more clearly by the gingers (Zingiberaceae) and arrowroots (Marantaceae) in the alliance of families
(Scitamineae) to which the banana belongs with its pachycaulous

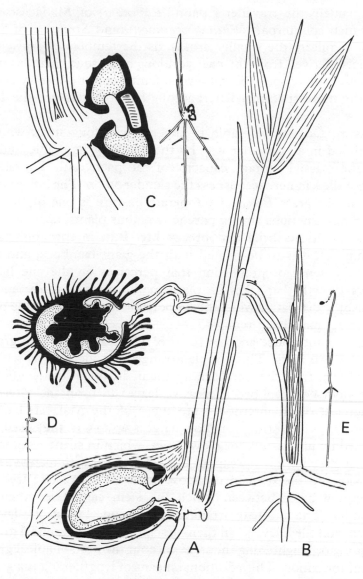

Figure 53. Monocotyledonous seedlings with the food store (endosperm) protected by the hard seed coat or, in palms, by the woody stone of the fruit (as shown in black); × ½. A, Nipa palm, with many scale leaves before the first foliage leaf. B, Forest palm *Pholidocarpus*, planting its seedling at a distance by lengthening the stalk of the cotyledon; the fruit stone with spiky fibres into the pulp (eaten off), and a woody flange into the endosperm. C, Banana seedling, and an enlargement (× 2) to show its water store (hatched). D, Rice seedling. E, Onion seedling, raising its small seed on the slender cotyledon.

157

allies, namely the traveller's palm (*Ravenala*) of Madagascar, its Amazonian counterpart (*Phenakospermum*), and *Strelitzia* of South Africa. Similarly the smaller aroids, as the temperate *Arum*, must have been derived from the pachycaulous because in them, as in the other cases, more specialized flowers, fruits, and life-history accompany the more specialized herbaceous habit that is not the forest factor.

Grasses, too, are referable to the pachycaulous state, which is exemplified in a particular way by the bamboo; its edible bud displays the massive storage capacity of the pachycaul. The slender grass parallels in herbaceousness the slender onion. The inflorescence of the quake grass *Briza* of temperate pasture is one of the most slender constructions among parenchymatous plants, apart from the mosses, as fits its brief and breezy life. But, in the grass family (Gramineae), it is to be related with the giant bamboos, and there must be a sense of proportion that perceives on the one hand a massive growth that takes its place in the creative forest and the diminishing derivative which fits opportune niches in wayside places where trees have been removed.

Bamboos introduce another feature. The leaves, as in all grasses, are set in two rows. This simple arrangement, derived as with the willow leaves from the spiral arrangement, facilitates the lengthening of the stem between two successive leaves to form the internode. The longest of all internodes are those of the Malayan bamboo, *Bambusa wrayi*; in straight hollow lengths up to 9 ft. they used to be trafficked by jungle-folk for blow-pipes, which in silent accuracy are the most effective forest weapon. Large leaves set with many vascular connexions to the stem in the close spiral of the rosette tree have overlapping bases between which the stem cannot lengthen into internodes. The alternate arrangement has one leaf at one level on the stem and these levels can then be separated by internodal growth, elongating or heightening the stem without the need of apical growth and leaf formation. This additional means of lengthening is a general feature of leptocaul construction and assists in the upward struggle to the light, but the evolution of internodes is another unwritten chapter.

A bamboo shoot begins as a massive bud at ground level. By lengthening the internodes, it rises rapidly, even as much as 10 in. in twenty-four hours, and thus may reach in the course of a few months to a height of 50–100 ft. in rivalry with trees. But, in heightening, the stem thins and begins to branch. The branches bear leaves and, in branching again, the branchlets and leaves become smaller until a minimum size is reached beyond which that kind of bamboo cannot grow. The bamboo pole begins as a pachycaul stem with very reduced scale-leaves (as bud-scales): it ends with profuse branching into slender leafy twigs; thus it shows how trees were formed.

We come upon a matter that has escaped attention in the consideration of trees; it is the all-pervading relation in land plants between the size or massiveness of the primary stem and its ability to branch. The more slender is the stem the more freely does it branch; the thicker the stem the less is it branched and the more pole-like its appearance; the thickest palms are unbranched. The explanation seems to lie, partly at least, in the action of growth substances formed at the stem-apex, and which, passing down the stem, inhibit the outgrowth, perhaps even the formation, of buds in the leaf-axils [57]. If a stem is cut off there develop from below the cut end buds that were dormant – that is, inhibited; the removal of the stem-apex, even the last inch, is enough to permit these buds to sprout. It seems that the bigger the stem-apex the greater is this chemical prevention of branching; therefore pachycaul trees are not only less branched, but the branches that do form arise at a great distance from the stem-apex. The explanation is by no means complete because cutting the crown off a palm will not enable it to branch. Nevertheless, the suppression of axillary branching by apical dominance, which we noticed in the growth forms of seaweeds, does seem to become a key factor in the growth form of trees. It appears indeed that the success and variety of flowering plants has accrued very largely from their exploitation of the internal control of growth by chemical means, fitting the general internalization of the land plant. This subject of growth substances at last knits growth with form; though in its infancy as a growing point of knowledge, it is beginning to make itself felt in many walks of botany and agriculture.

The way in which buds are suppressed helps to give the tree its characteristic shape. If all buds developed from all the leaf axils, the result would be a thicket of competitive twigs and an enormous waste of energy. In fact extremely few buds are grown. Thus the pine tree of elementary botany acquires unusual interest because it is one of the very few kinds of tree which, differentiating short shoots of limited growth from long shoots that will become branches, manages to grow every bud on its long shoots except for those of the first few seedling leaves.

From the centre of its crown of leaves, the palm thrusts a sword into the air. It opens to unfold the leaflets from above downwards, expressing the way that the leaf has grown from the base and pushed the oldest parts out of the bud first [58]. It is the way of all mono-cotyledonous leaves, and even that of pine needles. The tender young parts are protected by the old leaf bases. Even so, knowing animals, such as bears and monkeys, may tug out with their teeth the young leaf of pachycaul monocotyledons and eat the succulent base. Con-trast the leaves of ferns, tree ferns, and many dicotyledons which grow at the exposed tips. They offer tender growth for browsing but, when eaten, the growth of the leaf is destroyed. Many dicotyledons, even potatoes and fig trees, have developed the protection of basal growth more or less, but it reaches its highest expression in the leptocaul meadows, grasslands, and steppes where hoofed herds graze without damage to the growing points; it is the principle, too, of the mowing machine. It is a property that must have contributed greatly to the survival of monocotyledons, whether in places of drought, frost, or over-grazing.

In displaying pachycauly as the primitive forest condition and leptocauly as the derived condition that leads to lower vegetation outside the forest, particularly in exacting temperate climates, the monocotyledons not only support the evidence of dicotyledons for the manner of forest evolution but emphasize the primitive nature of the pachycaul. Palms exist now throughout the tropics in great variety with roughly two hundred genera and three thousand species. Many are highly peculiar with most specialized inflorescences and fruits, and not a few, like the double coconut (*Lodoicea*), which occurs only on two small islands of the Seychelles group, show also in their geographical distribution evidence of great age. By contrast the grasses, sedges, and rushes display modernity in construction, their habitats outside the forest, and their wide distribution. The home of palms, screw pines, banana-allies, and other pachycaul monocotyledons is the swampy flood zone of tropical rivers, where on newly forming soil they seem to show how the flowering forest was pioneered.

Chapter 10

The Seed

MECHANISMS wear out. Living mechanisms, by which expression scientists subject living beings to their arguments, prepare for their replacement. What the plants acquired individually during their lives they return, under the benthic law of survival (p. 80), as offspring for the benefit of their race. Cells co-operate to make a plant structure capable of dealing with the environment, but it wears out. It reproduces in the seaway by spores and gametes. Spores persist in, and gametes having conjugated revert to, unicellular selfishness. Remembering in their chromosomal heritage how to grow, they start the new generation in competition; the heartless victors make the new plant structures that repeat the life-history. Ethics, however sublimated in the animal world, belongs with protoplasm and in vegetable simplicity pervades plant life. Enormous, compared with those of the sea, are plant structures on land. Enormously they wear out, and enormously they have transformed the sea reproduction that they inherited into the seeds of the land. Coconuts bob where seaweed spores prevailed.

Seeds, trafficked as little boxes with a plant inside and the subject of many a parable, are commonplace. Their ease is the art of nature, but seed-making in the first place involved so profound a reorganization of the plant's life-cycle that its history is peculiarly botanical. The full story is very far from known, but there is enough to reveal what seeds have come from and why they brought success in varying measure to the seed ferns, the cycads, the conifers, and the flowering plants that replaced the early forests of seedless ferns and their allies. Though they are made from a sea recipe, there is nothing comparable with a seed in the lives of the seaweeds. Of the four great inventions of land plants, namely internal air spaces between the cells, cuticle, lignin, and seeds, they are geologically the latest and by far the most complicated [59].

A seed must lodge somewhere for some time in order that the young plant within may emerge and root. Such lodgements cannot be made on the rocks in the moving water of seaweed life, though sea grasses and mangrove trees bring their seed heritage from land and establish a new order of seed sea plants on muddy and sandy flats. The fact is that no seaweed has had to take even the first step in seed evolution, which is the transference of sexuality from gametophyte to sporophyte to produce dissimilar spores comparable with dissimilar gametes. Nevertheless, a knowledge of sea life-cycles is as necessary as a knowledge of sea structures, to understand the lives of land plants.

Mosses and ferns reproduce by spores. A spore is a single microscopic cell strengthened by a thick wall and waterproofed against evaporation by firm cuticle. The spore with cuticle is the mark of a land plant: seaweed spores are surrounded by a film of watery mucilage. The brown powder from the underside of fern leaves and the dust within a moss capsule are the massed effect of thousands, if not millions, of spores. As dust they blow about and settle to grow in suitable damp places into new plants. The naked eye cannot see the start of a moss or fern, and days, even weeks, must pass before the microscopic structure becomes tangible. Here is the advantage of the seed. It starts with a big mass of cells, many stored with food reserves (starch, oil, protein) for the growth of the young plant within, which is so well formed that, as soon as the seed has lodged, even without having to absorb water, it can put forth root and leaves and at once take station and effect. Seedlings begin to manipulate the environment and to overshadow 'sporelings' almost before these have got going. Bigger seeds will produce bigger seedlings to outstrip the smaller, and there started long ago another field of competition in the botanical preparation for space travel [60]. Nuts are champions, among which palm seeds are the prize-winners: they can carry enough food to provide a seedling palm with several leaves, up to a foot long, and many strong roots. There are trees and lianes in the forest, too, with big seeds that will send a leptocaul shoot 4 or 5 ft. into the air and the root as deeply into the ground. By the mid-Palaeozoic era seeds had already gained the ascendancy over spores as the means of tree reproduction, and seed plants had begun to replace the tree clubmosses, horsetails, and ferns.

Spores persist, nevertheless, because their minuteness enables them to be produced in vast numbers sufficient to find out by chance arrival the multitude of microscopic places on the earth suitable for plant growth, but where seeds cannot enter. The crannies of rocks, trunks, branches, leaves, mounds of earth, declivities and such like provide innumerable places: ferns and mosses may even grow on the coconut husk. More lately in geological time the forest seed itself

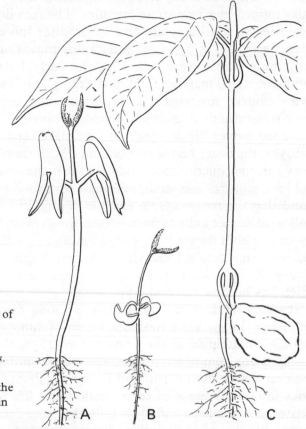

Figure 54. Seedling variation in a family of forest trees (Dipterocarpaceae); × ½. A, *Balanocarpus*. B, *Shorea*. C, *Dipterocarpus*, with the cotyledons retained in the nut.

diminished in size and travelling lightly began to exploit the same places. Acorn and chestnut sprout where they fall or roll. Birch seed and thistledown are blown long distances. Orchid seeds alight on trees. Alder seeds drift on the water. Fig seeds travel in the insides of animals, grass seeds in their fur. The variety evolved by the seed plant is picked up by some factor in the environment, reflected to the plant, returned to the environment until, to and fro, there echoes down the corridor of time a babel of instruction how to despatch a modern seed [61].

When a spore of a moss or that of a liverwort grows, it passes through a filamentous stage and, in seaweed manner, builds another moss or liverwort. But the fern spore grows into a small liverwort-like plant, a few millimetres wide, called the prothallus (or previous thallus), from the underside of which sprouts a little fern. The moss or liverwort, on reaching full growth, makes then its upright, leafless stalks topped by single spore capsules. The fern develops bigger and bigger leaves until they are able to produce spores on their undersides. Because they appeared to lack any means of sexual reproduction, these plants were called by the early botanists cryptogams ('with concealed marriage'). Higher seed plants, called phanerogams ('with evident marriage'), had stamens and ovaries, which were easily recognizable as male and female equivalents because the ovary contained ovules ('little eggs') and, to make seeds out of them with embryos, the ovary had to be pollinated with staminal dust. Now we know through microscopic study that it is the so-called cryptogams that have gametes and straightforward sexuality, because it has been found that the moss capsule and the young fern plant (not its prothallus) arise, not as buds from pre-existing tissue, but from fertilized egg-cells within the leafy apex of the moss and within the prothallus; and with this detailed knowledge botany began to reconstruct the history of the seed. It is the phanerogam or seed plant, however, which has become the real cryptogam.

'It was to Hofmeister, working as a young man, an amateur and enthusiast, in the early morning hours of summer months, before business, at Leipzig in the years before 1851, that the vision first appeared of a common type of Life-Cycle, running through Mosses and Ferns to Gymnosperms and Flowering Plants, linking the whole series into one scheme of reproduction and life-history' [62]. Hofmeister gave the facts and Darwin in the same decade their explanation by his theory of evolution. Botany strode forward.

The fern prothallus is a small parenchymatous mass of photosynthetic cells. From the underside grow fine root-hairs into the soil. It lacks stomata, lignification, vascular bundles, and any differentiation into stem, leaf, or root. On the underside it bears microscopic male and female sex organs, similar to the gametangia of seaweeds but provided with a sterile, protective, layer of cells. The male organs, called antheridia, produce larger numbers of sperms (antherozoids of botany) set with many flagella by which they swim in films of water between the prothallus and the soil to the female organs, which are called archegonia. An archegonium contains but one egg, which is, like the egg of the red seaweed, presumably the result of reduction from a larger number. The egg is not set free but waits at the bottom of the flask-shaped archegonium for the sperm, which

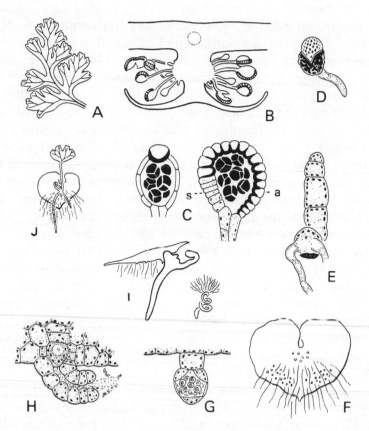

Figure 55. Features in the life of a fern. A, Part of a leaf to show the formation of the pinnate frond by dichotomy with overtopping; ×2. B, Section through a leaf to show the group (sorus) of sporangia under a cover (indusium), supplied with a vein; ×15. C, Sporangia (cut in two directions) with the tetraspores; the ring of thick-walled cells (a, annulus) causes, on drying, the ripe sporangium to split at the thin-walled part (s, stomium); ×60. D, E, Spore (with thick dark cuticle), growing into a short filament; ×300. F, Prothallus produced by the filament, bearing root-hairs and sexual organs on the underside; ×6. G, Antheridium forming many sperms (antherozoids); ×300; a coiled sperm with many flagella; ×900. H, Archegonium with one egg-cell, with its neck opened and attracting sperms; ×300. I, Young fern plant growing from the zygote on the underside of the prothallus (in section); ×6. J, Young fern plant with its first leaf and roots; ×3. (A, after Bower, 1923; B–J after Kny, 1894.)

enters through the mouth at its apex. When the mouth opens a row of sterile cells (the sterile eggs, presumably) breaks down and makes a slimy track that stimulates the sperms to draw near and allows one to swim in and conjugate with the egg. The zygote is formed and it starts the new generation that grows into the new fern plant; with leaf, stem, and root, it soon establishes itself independently and smothers the short-lived prothallus. The new fern plant is often, but wrongly, called the sporeling, following the old ideas, but the prothallus, developed from the spore, is the sporeling.

The fern plant is the diploid sporophyte. The prothallus is the haploid gametophyte. The zygote commences the new diploid generation and meiosis (halving of the chromosome number) occurs immediately before the spores are made. The spores are borne in minute sporangia set in clusters (sori) on the underside of the frond, or, in some ferns, on the edge. The sporangium has an outer sterile layer of cells forming a protective wall, as in the gametangia and as the epidermis of the leaf. Inside are many cells that act as spore mother-cells by undergoing meiosis and forming groups of tetraspores. When they are ripe, the sporangium wall dries, splits, curves back the end part that carries the spores, brings it sharply forward again, and slings out the spores, which scatter in the air. For the intricate details of this and other mechanisms of spore dispersal in land plants, I refer to the handy volume by Ingold [63].

The fact is that every spore mother-cell in the fern sporangium is a tetrasporangium, equivalent to those of *Dictyota* and the red seaweeds. The many tetrasporangia are aggregated into a sorus set in a special outgrowth from the surface of the leaf, which is called the sporangium but, as a sorus of tetrasporangia, it is a super-sporangium. Then these super-sporangia are set in a super-sorus which is the sorus of the fern leaf. The names were invented before the structures of marine botany had been correlated with those of land plants; hence the confusion. The marine term for the fern sporangia would be a stichidium, and though red seaweeds, accustomed to deal in tetraspores, go in for stichidia, the word and even the notion is unfamiliar in land botany. Nevertheless, the fern sporangium appears as a marine structure with cuticle, epidermis, internalization of mother-cells, and air-dehiscence for airborne spores, fitting its land life. It exemplifies the way in which land plants turn the small reproductive structures that they inherited into beautiful little air mechanisms, almost toy-like, among which the moss capsule is a favourite. Fungi, as we shall see in Chapter 13, turn the whole reproductive plant body into toy umbrellas, brackets, pepper-pots, and such-like. What in the sea was both growing and reproductive must be refitted on land to make a reproductive mechanism fed by the growing or vegetative part of the plant; hence leaves precede flowers, and flowers are more often parasitic than seaweeds.

The fern life-cycle resembles that of *Laminaria*, but tetraspores take the place of zoospores, the gametophyte is less reduced, and the egg is fertilized *in situ*. The big sporophyte, fitted with roots, takes over the large output of spores for dispersal and re-establishment, with the inevitable high wastage of a random method. The spore grows into a small, rootless prothallus that nestles in wet places suitable for the free movement of sperms in their primitive aquatic manner outside the parent body. The egg, retained in its oogonium (turned into the special archegonium, or walled structure, of the land) draws on the prothallus after fertilization for the food supply to enable it to develop into the young sporophyte: the one egg avoids immediate strife, though the prothallus bears several to many archegonia among which only one succeeds in establishing a new plant. Thus the fern shows in remarkable parallel with modern sea-weeds the adaptation of its seaweed ancestry to land. The delicate prothallus that was a sea success in quiet water is now an impediment. It needs a wet place in which to grow and to function sexually. So, by ditches, banks, waterfalls and drips, on misty mountains, and in all the dankness of wet undergrowth ferns are obliged to prosper. The prothallus is the Achilles' heel of the land sporophyte. As a diminutive parenchymatous structure, practically undifferentiated save into sex organs, it was never the primary form of a seaweed but must have been simplified already in the seaweed ancestor of the fern into a reproductive piece of a marine life-cycle [64].

The moss and the liverwort have also the heteromorphic life-cycle but with the very great difference that the moss or liverwort of ordinary recognition is the haploid gametophyte and the thread with its spore capsule is the diploid sporophyte. It grew from an egg in an archegonium at the end of the moss stem or on some part of the liverwort. The little snuffer at the tip of the young moss sporophyte is indeed the wall of the archegonium enlarged after fertilization, ruptured by the growth of the sporophyte, and serving as a protection to its tip. Neither gametophyte nor sporophyte has roots. The gametophyte is limited to the same situations as the fern prothallus, though mosses have acquired remarkable powers of drying and can survive on rocks, open ground, twigs, and other places that are without rain for a long time. The sporophyte cannot get off the gametophyte, where captured like the carposporophyte of the red seaweed it becomes a spore mechanism, which its lack of leaves facilitates. Its capacity to fill the environment is very small; its output of spores is limited; it can in no way become a major element of land vegetation.

The moss and the liverwort are curiosities. With relatively large gametophyte and small, cylindrical, and leafless epiphytic sporophyte, there is nothing like them among modern seaweeds. Yet a sea counterpart may not have been impossible. If fertilization *in situ* were part of the property of a parenchymatous green seaweed and it formed a carposporophyte that had tetraspores, there would have been a beginning. But here lies the interest of the ancient, lower Palaeozoic fossils known as *Rhynia* and *Hornea*, which are so well described in all books on fossil plants [32], for they suggest a more elaborate and free-living state of the moss sporophyte from which modern mosses have been reduced. What is simple and works well now in a derived habitat is almost surely reduced.

A seed contains an embryo plant. This must have come from a zygote. The egg must have been in the reproductive tissue of the parent seed plant. Then, because there seems to be no other life-cycle inherited by green land plants but that in which sporophyte and gametophyte alternate, the gametophyte that bore the egg must also have been in the reproductive tissue, and, indeed, the spore from which the gametophyte grew could not have been liberated. Where is this spore? It must be haploid and therefore recognizable by microscopic study.

In the centre of a flower is the ovary. In the ovary are the ovules, remarkably like the young pale-green sporangia of ferns, though bigger. When the flower is pollinated, the ovules enlarge to become the seeds and in them is the embryo plant. The spore must be in the ovule. When the ovule first begins to form, as a minute hump from the inner wall of the ovary, there appears in it a cell recognizable from its denser contents and larger nucleus. This nucleus undergoes meiosis and there result four haploid cells as tetraspores, but they are never set free. The ovule never opens; if it did, the spores would merely fall into the ovary cavity. In fact, the spores are not prepared for a journey; they are not isolated and cutinized, but remain in the cellular tissue with unspecialized cell-walls, and, indeed, three of

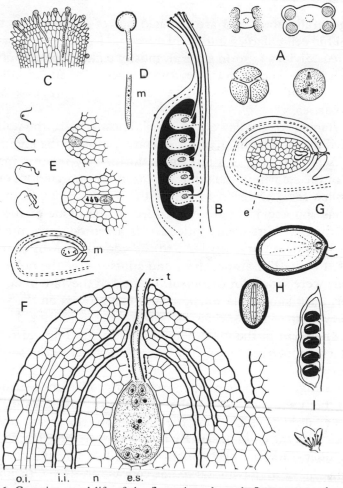

Figure 56. Cryptic sexual life of the flowering plant. A, Immature and an open anther in section, to show the four pollen-sacs; ×15; a pollen mother-cell (tetrasporangium) with the nuclei dividing at right angles in the tetrad manner to form tetraspores; ×250. B, Carpel in longitudinal section to show the ovules, each with a stalk, embryo sac, micropyle, and vein, and the pollen-tubes (exaggerated) grown from the pollen on the stigmas; ×30. C, Stigma with pollen-grains; ×120. D, Pollen-grain with its tube, containing two male nuclei and a vegetative (sterile) nucleus; ×250. E, Development of the ovule to show the growth of the two jackets (integuments) around the original apex (nucellus or megasporangium), ×50; two stages enlarged to show the megaspore mother-cell and the four haploid spores produced from it (three dying); ×350. F, Full-grown ovule (×90) and its apex (×350), to show the outer and inner integuments (o.i, i.i.), nucellus (n), micropyle (m), through which the pollen-tube entered the embryo sac (e.s.); eight haploid female nuclei and two male (black) nuclei in the embryo sac, the three central ones about to unite to form the endosperm nucleus; the egg is central in the upper three nuclei. G, Developing seed with the young diploid embryo growing into the triploid endosperm (e). H, Mature seeds with the embryo embedded in endosperm. I, Mature but unopened fruit (pod) and the flower from which it formed; ×½.

them generally abort and are crushed out of existence by the survivor. This cell enlarges into a microscopic lake of protoplasm, and its nucleus divides to form several, mostly eight, haploid nuclei, one of which with some attendant cytoplasm serves as an egg: the whole structure is called the embryo sac of the ovule, because in it the embryo arises.

The stamens of the flower, as is well known, make the pollen that is carried by wind, animal, or other conveyance, to the stigma of the ovary. A pollen-grain is a tetraspore formed by meiosis inside the anther of the stamen in the same way as in the sporangium of a fern. The pollen sacs, however, in which the pollen-grains form, do not project as sporangia but are embedded in the tissue of the anther. The stamen is short-lived; indeed it is a reproductive mechanism itself, which does not contribute to the general growth of the plant but is fed by it. The anther dries and splits to free the pollen-grains as if they were spores for dispersal. Dispersed they are in one way or another, but the only place where they can grow is on the stigma of the ovary of the same species of flower; therefore they do not distribute the plant in the inanimate environment. They distribute its hereditary properties to other flowers; they act as airborne male gametes.

On the stigma the pollen grain grows into a filament which thrusts its way between the cells of the style down to the ovary where, somehow, it finds entry into the ovule. Its tip seeks, finds, and coalesces with the embryo sac. By now there are three haploid nuclei in the pollen-tube, as this filament is called. Two enter the embryo sac while the third, having served as conductor, dies. One joins with the egg to form the zygote. The other joins with two more nuclei of the embryo sac and, united, they form a triploid nucleus (with three times the haploid chromosome number). The diploid zygote grows into the embryo, which forms a root and a seed leaf or two, but the triploid nucleus divides into a mass of triploid cells which become filled with food reserves; it builds the tissue, called the endosperm of the seed, on which the embryo draws for its food during its growth in the developing seed and, in many cases, during its germination from the seed after dispersal.

As embryo and endosperm enlarge, the wall of the ovule enlarges, and also the wall of the ovary enlarges to accommodate the one or more growing ovules inside it, which as many pollen-tubes have fertilized. The flower stalk thickens to supply with food and water the growing fruit with its developing seeds. All this is common knowledge of elementary botany, and it is becoming understood in physical and chemical terms how pollen-grains germinate on the particular stigma and how, by means of growth substances produced from the fertilized ovule, its wall, that of the ovary, the flower stalk, and any other part of the flower that enlarges or persists in the fruit, are stimulated into renewed growth after fertilization [57]. The un-pollinated flower normally drops off, as a reproductive mechanism that has failed. The problem is to discover what has happened in the life-history of the flowering plant; there are no gametes or gameto-phytes in any ordinary sense, and the plant is distributed by seeds instead of spores.

We turn to the cardinal points of meiosis and conjugation in the heteromorphic life-cycle. Pollen-sac and ovule can be recognized as super-sporangia producing tetraspores by meiosis, but the ovule has been reduced to the condition of one functional spore. There are plants that illustrate this egg-like reduction in numbers because they have variously two, three, or more spores in the ovule, where they produce as many embryo sacs, though only one succeeds eventually in the competition to form an embryo. I refer to the concise book by Maheshwari, where these matters and other details of the reproduction of flowering plants are explained [65].

Conjugation can be recognized in the union of the two haploid nuclei to form the diploid cell from which the embryo grows, but not in the triploid nucleus that forms merely sterile endosperm. The flowering plant is the diploid sporophyte but, in an entirely new manner to marine botany and to ferns and mosses, maleness and femaleness find expression in it as stamens and ovary. Sexual tetra-spores are produced; the male spore functions in a male way seeking the female tetraspore, which in the female way stays with the food. Both are now dependent on the parent sporophyte for their growth because they are not vegetative structures coping with the inanimate environment, but, like parasitic fungi, are supplied with food from the tissue of the ovary. Gametophytic tissue is reduced to the mini-mum of the pollen-tube or filament from the germinating tetraspore, and to the embryo sac from the stationary female spore; internal cell-walls and sex organs are dispensed with; the original male gamete, no longer discharged into the open to make its own way, becomes like the female a sexual nucleus.

By transferring maleness and femaleness, that is sexuality, from

171

the gametophyte to the sporophyte, the gametophyte part of the life-history is eliminated as a free-living state of the plant. It survives as the barest chromosomal memory required for the hereditary fulfilment of the life-cycle; each vestigial gametophyte, male as the pollen-tube, female as the embryo sac, forms but one sexual nucleus. What was a plant state living on its own to reproduce the race sexually becomes an inherited reproductive mechanism replacing the male gamete with a tube and structuralizing conjugation: for this reason seed plants are sometimes called by the quaint name *Siphonogamia* ('conjugating by a tube'). The Achilles' heel of the fern is overcome and the female super-sporangium becomes the seed, which is the new means of distributing an embryo sporophyte instead of by a haploid tetraspore; for this reason seed plants are also known as *Embryophyta* as well as *Spermatophyta* (meaning seed plant). The whole process shows the characteristic manner in which the successful plants of the land have converted their unsuitable sea inheritance into land mechanisms. The land plant, at first with dual and heteromorphic existence, transforms virtually into a fairly straightforward diploid individual viviparous as a mammal.

If this is not believed, then it must be realized that there are flowering plants that have proceeded further and eliminated even zygotes. If the pips of orange, lemon, or some other kind of cultivated *Citrus* are sown, or the flat stones of mangos, it will often be found that several seedlings grow from one seed. They are not the progeny of several zygotes in one ovule, nor are they cleavage twins of one zygote. They have grown each from a diploid cell of the sporangial (sporophytic) tissue bordering the embryo sac, which no longer functions. Nevertheless, pollination is still necessary to provoke the development of these asexual embryos and to form the seed about them. Because they are clones, or vegetative buds without sexual intervention, they are useful in agriculture for the propagation of pure lines. Freaks though they are, they are useful and from them will be learnt a great deal about the chemistry of pollination, fertilization, and embryo growth [66]. In the famous mango trick of India, the magician puts a mango stone into a pot, covers it with earth, and places a cloth over the whole. Then, as he pipes to it, the cloth writhes slowly up in the centre and, on its being removed, a leafy mango seedling (or is it the shoot of a twig?) is revealed. Botanical suspicion is roused because of one, not several seedlings, from the single stone.

There are two, apparently trivial, but novel details about the seed that must be mentioned because they have great consequences, as will be explained in Chapter 12. They are the triploid endosperm and the seed coat.

Haploid and diploid plants there are, but triploid are scarce because there is no normal way of making them and, when made, as in certain pollinations of flowering plants, they are usually sterile because three sets of chromosomes are difficult to divide satisfactorily into two viable or permissible sets at meiosis: one team is bound to get two centre-forwards. Triploid endosperm is a sterile tissue, which, as the larder for the embryo, resembles the last function of the prothallus. Some consider therefore that the triploid fusion is a means of reviving the memory of the degenerate female prothallus or embryo sac and causing it to develop into this state of the prothallus after fertilization. Recondite though it be, it is the tissue that makes up most of the cereal grain on which civilization is founded, and, for aught we know, is still the secret of the Corn Mother.

The ovule is a sporangial bulge from the ovary wall, comparable with a fern sporangium, but it is covered with one or two outgrowths from its attachment. They are called the integuments of the ovule because they grow round it like jerseys and draw together at the apex to form a small hole or neck, which is called the micropyle. The sporangial bulge that they contain is called the nucellus and inside it is the embryo sac. Through the micropyle the pollen-tube generally enters to reach the embryo sac. After fertilization the integuments increase in size, generally by much cell-division, to form the hard and often lignified seed coat or protective wall of the seed and the micropyle persists as the tiny hole through which water is absorbed for the revival of embryos that have entered the dormant state. The tissue of the nucellus, however, is absorbed by the endosperm as it grows and is usually no longer recognizable in the full-grown seed. Again the evolutionary origin and nature of the integuments is uncertain, but they introduce surprising effects into the problem of the fruit developed from the flower.

All these ideas about the origin of the seed of the flowering plant and the interpretation of its parts are confirmed by the conifers and cycads, which are less advanced seed plants [67]. They are surely not direct ancestors or even offshoots of the direct ancestry of the flowering plants, which is still in Darwin's words an abominable mystery, but they amplify steps that in the flowering plant may seem too speculative.

The cones of conifers produce either pollen or seed, never both. They represent maleness and femaleness transferred to the sporangial parts of the sporophytic tree, as stamens and ovaries in the flowering plant, though not combined in one 'flower'. The pollen-sacs lie on

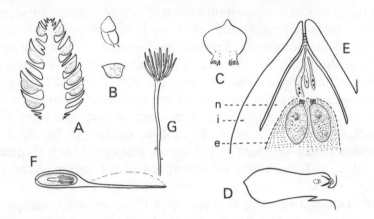

Figure 57. Reproductive details of the pine tree. A, Male cone in section, to show its die-away form; ×3. B, Male cone scale (microsporophyll) with two pollen-sacs on the underside; ×3. C, Female cone scale (megasporophyll) with two ovules on the upper side, each directed towards the cone axis and with a drop of fluid at the micropyle to catch the pollen-grains; ×3. D, Same as C in section to show the single integument; ×6. E, Apex of the ovule to show the integument (i) with micropyle, the nucellus or megasporangium (n), and pollen-tubes growing through to the archegonia in the endosperm (haploid gametophytic tissue); ×40. F. Ripe seed with the embryo embedded in haploid endosperm, the wing of the seed formed from a slip of the cone scale; ×2. G, Seedling with numerous cotyledons; ×2.

the underside of the scales of the male cone, but the ovules are exposed on the upperside of those of the female. Pollen is blown on to the micropyle, where it is caught in a drop of sugary liquid secreted from the ovule and, as this dries, the grains are drawn inwards to the nucellus, which they penetrate with short pollen-tubes. Inside the nucellus there is a fairly large mass of haploid cells, round the periphery of which there can be recognized archegonia. Pines have two to five archegonia in each ovule but the redwoods (*Sequoia*), known usually just as giant trees, have up to sixty. There is no mistaking the fact that this haploid structure, grown from a haploid cell in the ovule, is the equivalent of the fern prothallus. It is, however, colourless, because it is embedded in the parental tissue where the female spore was formed and kept: from this tissue the gametophyte draws its food and water supply. After fertilization the haploid tissue grows into the endosperm of the seed; there is no second fertilization to form, as in flowering plants, the triploid endosperm. As there is more than one archegonium in the prothallus, more than one zygote may start to make an embryo, but the embryos are rivals and only the most rapacious survives. Here is imperfection before the female gametophyte has been simplified to a single egg. Then, the pollen-grains form inside them one or two small cells that seem to play no part in the sexual process other than the beginning of the pollen-tube, and they are interpreted as vestiges of the male prothallus.

Cycads look like tree ferns. They have large, if leathery, pinnate leaves, but they bear large male and female cones, resembling giant pine cones, and their trunks, seldom branched, thicken slowly by cambial growth. They seem halfway between ferns and conifers. Their ovules are constructed much as in the conifers. The pollen-tubes, however, are very short and, instead of producing merely male nuclei to fertilize the eggs, out of the end of the tube swim slowly majestic sperms. Some cycads have merely two sperms from each pollen-tube, but they are 200–300μ wide and visible to the naked eye. The American *Microcycas*, which has one hundred to two hundred archegonia in the ovule, produces about twenty sperms, each 50μ wide, from the pollen-tube; as only one embryo survives in the seed, this genus shows well the waste of preparation required by the inherited sea recipe, before it is simplified to the single action of the flowering plant.

In the drop of liquid at the tip of the nucellus the sperms propel themselves with their many flagella, as in the fern, into the archegonia where one will conjugate with the egg. These stout plants of geological antiquity are unhurried. As trees they live for hundreds of years, even several thousand, and as sperms they may navigate the minute ocean which is made for them in prehistoric memory for one or two months: from pollination to conjugation four to six months may elapse. The eggs, also, are among the largest in the plant kingdom; that of the American *Dioon* reaches 6 mm. by 0·5 mm.

Cycas itself offers more evidence of an intermediate position between the fern and the conifer. It has no female cone but a cluster of

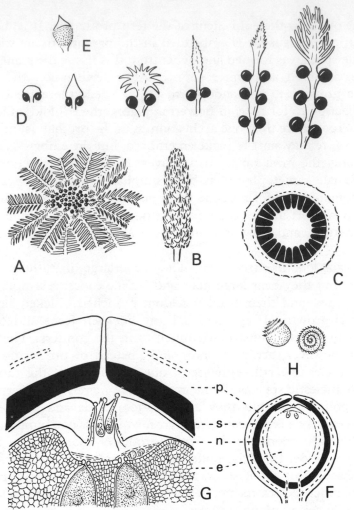

Figure 58. Cycad features, in their massiveness greatly reduced and in their minuteness enlarged. A, Centre of the crown of *Cycas* with the cluster of seed leaves (megasporophylls); $\times \frac{1}{40}$. B, More typical cone, male or female; $\times \frac{1}{40}$. C, Cross-cut of the stem with wide pith, wide rays through the secondary wood (in black), and wide cortex surrounded with persistent leaf bases; $\times \frac{1}{12}$. D, Series of seed leaves (megasporophylls) to show their reduction in size to cone scales with two seeds (as the two figures on the left, the remainder being seed leaves of species of *Cycas*); $\times \frac{1}{10}$. E, Male cone scale (microsporophyll) with many minute pollen-sacs on the underside; $\times \frac{1}{10}$. F, Ripe seed with the nucellus (megasporangium, n) containing the endosperm (haploid prothallus, e) with two egg-cells, and the integument divided into an inner woody stone (s), and an outer pulpy layer (p), the micropyle at the tip; $\times \frac{1}{2}$. G, Apex of the ripe seed showing the two archegonia with large eggs, the haploid tissue of the prothallus in which they are borne inside the nucellus, and the pollen-tubes producing sperms; $\times 250$. H, Two sperms with many flagella arranged in a spiral; $\times 500$.

short, imperfectly developed leaves, each carrying on either side one
to three or four large seeds. Furthermore, transitions can be found
between these reduced seed fronds and the normal pinnate leaves,
which show that the seeds take the place of the lower pinnae (or
leaflets). Aware that *Cycas*, in this respect, was a link with the fern
construction, but unable to study its reproduction in detail, Hofmeister guessed in 1863 that it would be found to possess motile
sperms. Nearly half a century later his guess was verified by the
Japanese botanist Ikeno (in 1898), but he was shortly forestalled in
this great discovery by his compatriot Hirasé, who reported in 1896
that the pollen-tubes of the maiden-hair tree (*Ginkgo*) produced
sperms. With foliage like a maiden-hair fern, but the upright habit
of a flowering tree with spreading branches, active cambium, and
hard, resistant timber, *Ginkgo* is the sole surviving species of another
ancient line of seed plants. *Cycas* and *Ginkgo* are two veteran ex-
ponents of the seed theory, explained in even greater excitement by
the fossil seed ferns [32]. Fossils, be it understood, are given botanical names and spoken of as if they were species, but they are merely
the bits, however wonderfully preserved, of extinct beings: we are apt
to forget in the explicit elegance of the structures made by cell-walls,
that it is the living protoplasm which is plant life and makes species.

Lastly, among living relics that help in the interpretation of the
seed, there are the clubmosses *Selaginella* (related to the Palaeozoic
trees *Lepidodendron*), the peculiar water quillworts *Isoetes*, and the
marsh or aquatic ferns *Marsilea*, *Pilularia*, *Azolla*, and *Salvinia* [68].
They do not have seeds but they show the first step towards the
seed habit, which is to possess distinctly different male and female
tetraspores. The male spores, called microspores because of their
small size, are produced in large numbers in microsporangia; on
being set free they form inside themselves microscopic male gametophytes, which are not photosynthetic but which produce a few
sperms to be discharged into the water. The female spores, called
megaspores because of their much greater size, are produced in
megasporangia, which contain either four megaspores (as in *Selaginella*) or only one as in the other genera, excepting *Isoetes*. The
megaspore is set free and on wet ground or in the water it forms a
small green prothallus, yet still microscopic, which bears a few
archegonia. The spores of *Selaginella* are dispersed by wind, but
those of the other genera are waterborne. In each case the sperm
swims its own way into the archegonium.

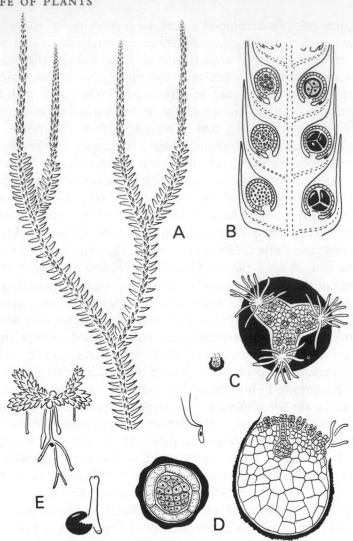

Figure 59. Life of *Selaginella*. A, Leafy sporophyte, dichotomous with two rows of small leaves and two of large, the stems ending in cones; natural size. B, Part of the cone in longitudinal section to show the microsporangia with many spores, the megasporangia (each with a single tetrad of spores), and the veins; × 12. C, Microspore and a megaspore germinated; × 70. D, Sections of these spores: microspore with central mass of cells, each forming one sperm (× 350), the megaspore with small haploid gametophyte bearing archegonia (one with an embryo sporophyte) and root-hairs (× 70); the sperm, × 300; note that the gametophytes are practically included within the spores. E, Young sporophytes, each with two cotyledons, growing from the gametophyte in the megaspore; × 30 and × 6. (C–E after Bruchmann, 1912.)

Thus in these plants, sexuality materializes on the formerly asexual sporophyte. Male and female sporangia produce tetraspores of their sex only and they grow into prothalli of the same sex. In transferring sexuality from the gametophyte to the sporophyte, there has crept in the same difference in size and number that distinguishes male from female gametes. There are many small male spores and few large female spores. Therefore these plants are called heterosporous in contrast with other ferns and fern allies such as *Lycopodium* and *Equisetum*, which now have to be called homosporous because their spores are not sexed but, alike in form and size, give rise to bisexual gametophytes that bear both male and female gametangia. *Isoetes* occupies a transitional state, as the most primitive indication of the seed habit in any living plant: its male sporangia contains about a million male spores 20–40μ wide, and the female one hundred to three hundred female spores 300–900μ wide.

Some kinds of *Selaginella*, of which there are about five hundred species, go a step further. The slender little cones bear the sporangia singly on the uppersides of the cone leaves. Some leaves bear microsporangia and are called microsporophylls; other leaves of the same cone, in some cases only the basal leaves, bear megasporangia and are called megasporophylls. The spores in most species are blown, tumbled, or flicked out of the sporangium, but in a few the female spores are too big for the inert female sporangium to liberate them. In this case the female spore produces its gametophyte (or female prothallus) inside the female sporangium, and its egg-cells are fertilized by sperms that swim in a film of rain water from male prothalli, produced in the male spores which have alighted on or near the female sporophyll, if not on the female sporangium. The embryo sporophyte then grows out from the female sporangium and eventually falls to the ground in a haphazard way, or indeed begins to grow on and over the parent; it has no means of dispersal.

Here then is the second step in the seed habit, which is the sticking or retention of the female spore in its sporangium where, in copying the female gamete, it has to work *in situ*. This arrangement in *Selaginella* nevertheless is still far from adequate to make a seed, because the female spore has a thick cuticle and it lies in a dried and dead sporangium. The female prothallus cannot draw nourishment for its growth from the parent plant, nor can the embryo that it produces: the food supply of the new generation is limited to that which can be stocked in the spore. It will be noted also that heterospory brings about the decrease in size and complexity of the gametophyte that has accompanied also the evolution of the seed. As the spores become the male and female units, the gametophyte is rendered unnecessary. The retention of the female spore illustrates, too, the process of evolution by failure of an inherited mechanism. The female sporangium fails to liberate the spore. If both sporangium and spore failed to mature their thick walls, and failed to ripen off, they would remain thin-walled and be able to draw on the sporophyll for further food supply to the embryo: this they might have done by neotenic progress, but it seems that they were unable.

Fossil remains give indubitable evidence that heterospory occurred also in the Palaeozoic *Lepidodendron* and *Calamites*, which may even have got as far as detaching the sporophyll with the female sporangium and spore inside, but still they seem not to have established the process of feeding the embryo via the sporophyll. As these plants seem to have had no direct connexion with cycads, conifers, or flowering plants, which probably had no direct connexion with each other, the conclusion is inevitable that in the past several kinds of fern and fern ally became heterosporous independently, and that some, only, of these succeeded, also independently, in becoming seed plants. In other words, the evolution of the seed is not the endowment of one excellent line of plants but a common goal variously attempted, just as all sorts of animals must have tried to fly but few, in different ways, succeeded. No one would classify a butterfly with a bat or a flying fish for this reason. Similarly, there are so many differences between *Selaginella*, *Cycas*, and a flowering plant that botany cannot classify them together just because they are heterosporous and on the seed route. For example, very profoundly does *Selaginella* differ from *Cycas*. The sperm of *Selaginella* has two flagella like that of the moss, liverwort, or *Lycopodium*. The sperm of *Cycas* has many flagella like that of the fern, horsetail, and *Isoetes*. If this difference occurred in seaweeds it would be fundamental, as a flagellate distinction: it probably is, in fact, in these land plants but the tenets of marine botany have not yet been assimilated in the arbitrary classification of land plants. As for the fossils, the nature of their sperms may never be known. How they and their survivors illuminate the evolution of the seed, perfected in the flowering plant, is therefore only by parallel instance.

If I were asked whether any living plant may be a relic of the line of flowering plants, I would point to *Marsilea* and *Pilularia*. They are

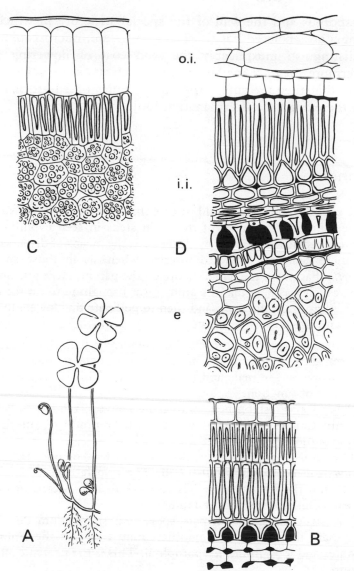

Figure 60. Significance of structural inheritance. A, Part of a plant of *Marsilea* with creeping stem and erect leaf composed of four leaflets, each leaf bearing two spore-boxes near the base of the stalk; × ½. B, Structure of the outer part of the wall of the spore-box, to show the columnar cells and 'hour-glass' cells (with air spaces, in black, between them); × 200. C, Outer part of the seed coat of the paeony, with starch grains in the cells below the columnar layer; × 100. D, Seed coat of the arnatto (*Bixa*) in section, to show the columnar and hour-glass cells in the inner integument (i.i.); the outer integument (o.i.) pulpy; the endosperm (e) with large starch grains; × 200.

heterosporous and the wall of the special structure in which their sporangia are enclosed has those very intricate, microscopic details that distinguish inexplicably the seed coat of flowering plants. Furthermore they are marsh or aquatic plants, so strange in many features that they are surely the merest relics of a host of ferns of which there is no fossil explanation [69].

Appendix

For the sake of those who would pursue this most complicated subject, I give the following summary of the main steps in seed evolution. This great event replaced the spore by the seed as the means of dispersal. In so doing there came an awkward interval when, as in those species of *Selaginella* that retain the female spore on the parent, there was no means of dispersal. It was here that the animal was brought in to eat the outside of the seed, to work the fruit, and then to pollinate the flower: to evolve, that is, with the flowering forest.

1. In the fern type of life-cycle (heteromorphic with dominant sporophyte) sexuality is referred from the gametophyte to the sporophyte, where it works out in male and female duality. This is evolution by the transference of function.

2. Heterospory established, the gametophytes reduce in size and photosynthetic activity. They become inherited reproductive mechanisms. This is evolution by simplification.

3. The female sporangium works down to few, or one, large female spore that can no longer be liberated. The transference of sexuality is complete and the female spore has to work *in situ*. Means of dispersal disappear. This is evolution by failure.

4. The structure of the female spore and sporangium deteriorates. They remain thin-walled and undifferentiated so that the embryo can draw water and food from the sporophyll. This is the neotenic answer to the failure.

5. The male prothallus becomes a pollen-tube, as a neotenic simplification.

6. The number of archegonia in the female prothallus reduces to one.

7. Spores and eggs disappear and their function is taken over by male and female nuclei: evolution by failure.

8. The tissues of the male and female prothalli are cut down until even cell-walls are not formed and there is free nuclear division in the pollen-tube and embryo sac: evolution by failure.

9. Triploid endosperm is introduced. The evolutionary history is unknown.

10. The female sporangium is invested with extra coverings (integuments) that make the seed coat. The evolutionary history is unknown, but possibly by transference of function.

11. The seed with its embryo is detached and restores the means of dispersal. The whole failure of dispersal is finally overcome and the seed plant begins to travel.

Chapter 11

The Flowering Tree

FROM THE EVIDENCE of the fossil record of plants and from the innate history of the seed, it is clear that early Palaeozoic forests were built by spore-produced trees. By the mid-Palaeozoic – and it is necessary to repeat that these animal eras do not correspond with botanical history – the spore forests were being substituted by seed forests, and this substitution has continued until nowadays the spore trees have either disappeared or are represented by such relics as the tree ferns or by herbaceous forms such as most living ferns, lycopods, *Selaginella*, or horsetails. In its wake the seed brought pollination, which is the transfer of male spores to the immediate neighbourhood of the female spores localized on the parent. Instead of scattering and losing in vast numbers spores at random, one kind is conveyed to the other, and about this instance of biological purpose, the flower has been evolved. It fits a need and, with function perfected, it has diversified into masterly variety that brought a new era into plant life.

Early seed forests gave place to flower forests, but no one knows when this happened or indeed when any major botanical transition took place; that is why geological history cannot be subdivided into botanical eras. Equipped with flowers, the new seed plants spread from the forests to the polar tundra and invaded freshwaters; they returned even to the sea, which no moss, liverwort, fern, fern ally, cycad, or conifer had been able to do. Emancipated, they transformed the vast gloomy flowerless world with gaiety, feast, and song into a new factory of life.

Over all the warmer parts of the earth, particularly in the rainy tropics where there are no adverse circumstances for plant growth, flowering plants predominate. They make the trees, the climbers, the epiphytes, the shrubs, and the herbs in variety far exceeding that which the flowerless seem to have accomplished. A single family, such as the Leguminosae, with some fourteen thousand living species, can outnumber all the surviving ferns (nine thousand species). The general structure of the seed has been standardized but it is borne on all sorts of plant growths, which have varied by branching, leafing, flowering, and fruiting into the modern spectrum of some three hundred thousand species. At least ten times this number if not a hundred times, must be added to fill the gaps of extinction. These developments attracted animals. The flowering forests began to feed them, to use them, to foster them, and to bring forth their counter-part in the multitude of insects. For the first time indeed in the history of life, its two arms began to co-operate. Animals do not help plant plankton, seaweeds, mosses, ferns, or conifers; they help themselves to them. The flower symbolizes the new community and indeed floreat! Until the tissues of the flowering plant had evolved, vegetable growths, as opposed to the microscopic plankton, seem not to have been appetizing; even dugong and manatee, returned to marine life, prefer seagrasses to seaweeds. On this fact the flowering plant has developed the strategy that employs animals in its prime needs of pollination and seed dispersal, and in the processing of vegetable excess into manure.

A flower is a reproductive bud. It opens not to form leaves where-by to add substance and reserves to the plant, but to spend these in the provision of offspring. Horticulture has so developed and trans-formed flowers into sterile double beauties of long life that it is easy to forget that the real function of the wild flower is a prelude, as brief as possible, to the production of seeds. Buds are a feature of land plants that adds to their precision. Part of the vagueness of seaweeds lies in the lack of buds; their branches, leaves, and lobes are the continuation of active growth as it exposes new surfaces under water. The delicate young leaves and branches of the land plant cannot be thrust at once into the air any more than the delicate growing points can be exposed: they would dry up. They are internalized within the protection of the older leaves; only at an advanced state of develop-ment, usually when cell enlargement by absorption of water is all that is needed to bring them to full size and activity, do they begin to emerge and function. Up till this time the young structures have been fed in the bud from food reserves in the stem. Thus the vegetative bud has already the principle of feeding that the repro-ductive bud exploits.

In some plants the buds consist entirely of foliage leaves, the outer and adult protecting the inner and younger. They are buds of continuous growth, though they may be checked for some time under adverse climatic conditions. Most trees, however, have small bud

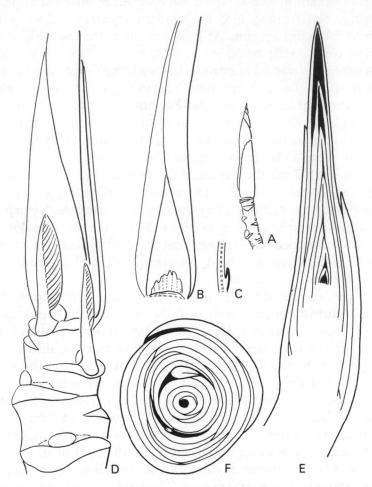

Figure 61. Bud development in a tropical rain-forest tree, the Malayan spindle tree (*Kurrimia*), allied with the European spindle tree (*Euonymus*). A, Bud covered by scales (stipules) at the end of the short flowering part of the twig, marked by scars of bud scales and inflorescences, the lower leafy part of the twig not shown; ×1½. B, Bud scale with the gland at its base that secretes watery slime to lubricate the extension of the young leaves; ×7. C, Scale in section to show the gland with thick secreting epidermis; ×7. D, Bud A with the lower three sets of bud scales removed to show the young leaves; ×7. E, Similar bud in section to show the protected stem-apex; ×7. F, Bud in cross-section, showing two young leaves; ×14.

scales that cover and protect the contents of the bud when it seems to be quiescent. Such are the winter buds of temperate trees, but they occur also in most tropical trees, the growth of which, for one reason or another, is intermittent [70]. Though evergreen these trees are not everleafing. Their buds open and develop new shoots rapidly, so that this tender stage is hastened, and then the buds relapse into a state of apparent dormancy when they form within them the next set of leaves. Bud scales are at once the beginnings and the vestiges of leaves; they start as leaves at the stem-apex, but their

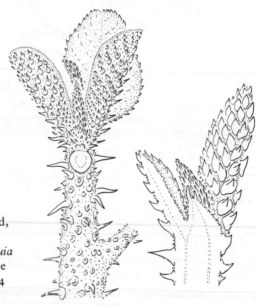

Figure 62. Heavily armoured, thorny twig and leaves of a small rain-forest tree *Saurauia* (New Guinea); × 2 (with the older leaves cut off), and × 4 (in section).

apical growth is soon arrested and they become tightly fitting, overlapping leaf bases without blades.

The immediate modification of a vegetative bud into a reproductive bud consists in converting the inner bud scales into fertile scales carrying pollen-sacs or ovules. The bud then opens to form a cone, the size of which depends on the number of fertile bud scales (sporophylls) that are formed, but in every case it will be found that the growth of this bud is limited to the one cone; it is a die-away structure fitting a certain food supply, tapering and therefore conical. Its size depends on the stoutness of the vegetative twig that produced it. Thus the massive pachycaul stems of cycads produce the

187

biggest cones, slender junipers and cypresses the smallest. Undoubtedly some physiological change has converted the vegetative or foliage bud into the reproductive cone- or flower-bud. In many massive plants a long period of vegetation continues for many years before reproduction begins; then it may continue uninterrupted during the life of the plant, as in many palms, or it may be seasonal, or indeed it may terminate the life of the plant. In other cases external factors such as dry weather, temperature change, intense light,

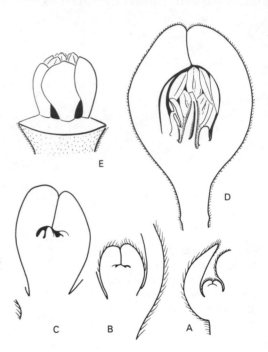

Figure 63. Bud protection of flowers in the leguminous tree, the flame of the forest (*Delonix*). A, B, and C, Sections of very young buds protected by the bracts and forming sepals in A, petals in B, and stamen and ovary in C; ×25. D, Half-grown bud in section with petals, stamens, and ovary well formed within the thick calyx; ×7. E, Similar bud with the calyx cut off; ×7.

or day length (that is the ratio of daylight hours to night hours), may cause flowering even after a very short period of vegetative growth. Why the land plant should enter the reproductive state is an immense subject, which joins now with that of growth substances; some of these can induce the formation of reproductive structures. It may be that the external factors affect mainly the production of growth substances, which thus appear as the internal cause, but it is yet impossible to generalize.

A flower is a highly diversified reproductive bud in which the scales are of five kinds, developed in succession with the inner two acting as the fertile male and female scales. The outermost scale or scales form the bracts: they are commonly the most leaf-like, or least reduced. Then come more specialized sterile scales that form the perianth of the flower and are usually distinguishable as the outer green sepals (forming the calyx), and the inner coloured ephemeral petals (forming the corolla). Internal to the perianth and later formed are the male scales or stamens and, in the centre of the flower, the female scales or carpels. The male and female scales are

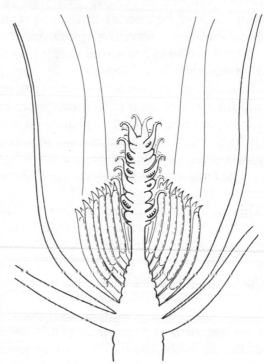

Figure 64. Section of the lower part of the magnolia flower of *Michelia chempaka*, to show the primitive form of the long flower axis, the long anthers, and the numerous free carpels; ×2.

equivalent to sporophylls and are, technically, the microsporophylls and the megasporophylls. All these modified scales arise in the leaf manner of phyllotaxy (or 'leaf order') on the stem-apex of the bud, but they are not usually spaced apart by internodes, except that the bracts are often separated by an internode from the rest of the flower, which is, accordingly, stalked. Hence, even when the flower-bud has opened, it keeps its compact form. In the paeony transitions commonly occur between leaves, bracts, sepals, and petals, and in the water-lilies there may be transitions between petals and stamens;

indeed, in the canna flowers, all the stamens may be red and petal-like except one, which has a small anther attached to one side. Generally, however, the distinction between stamens and carpels is sharp: in monstrous or abnormal flowers extraordinary intermediates may appear. The flower, it will be noted, is also a die-away structure and therefore cone-like [71].

Then, besides these modified flower leaves, there are the nectaries, which are patches of epidermal glandular tissue secreting the sugary nectar, situated in various parts of the flower. They are not the equivalent of bud scales but are specially inserted vestiges of the glandular, sugar-secreting epidermis of the land plant, which finds expression also in the so-called extra floral nectaries on various parts of stems and leaves, especially in tropical plants. The significance of extra-floral nectaries is obscure; they undoubtedly attract ants, and they will undoubtedly attract more botanical attention when the very close relation between plants and animals in the tropics is better realized, and when we have some idea of epidermal evolution in land plants [72].

The female part of the flower, called the ovary, encloses the ovules, and pollen, in consequence, is not conveyed directly on to them. This is the great difference from the female cones of cycads and conifers, in all of which the ovules are exposed and pollen is deposited on the micropyles. For this reason these plants are called gymnosperms ('naked seeds', 'naked ovules') in contrast with flowering plants called angiosperms ('with seeds, and ovules, boxed in').

Ovular boxing takes three forms. In the simplest condition, the female bud scales or carpels are folded longitudinally with the two sides apposed, as happens in many young leaves. On the edges the ovules are borne alternately so that they fit into a single row between the two sides of each scale: they are hidden by nestling between the two sides of an unopened young leaf, modified into a bud scale, and the tip of this leaf becomes the place where the pollen-grains arrive and proceed to grow towards the ovules. The tip is called the stigma and if, as often happens, the tissue below it lengthens, it is supported on a short column, called the style, leading to the body of the structure containing the ovules.

The pollen-grain on the stigma, as mentioned on p. 170, grows by a tube down the ovary tissue to the ovule. All sorts of pollen from different species of plant may alight on the stigma but, for chemical and physical reasons, most of the foreign pollen cannot force an entry: thus the style acts as a filter. In recent years it has been found that pollen from the same flower or individual plant often grows more slowly on the style than pollen from another plant of the same species; if cross-pollinated, the cross-pollen tubes reach the ovules before the self-pollen tubes, which may arrive and fertilize ovules only in the absence of cross-pollen. Thus the style is not only a safeguard against foreign pollen but a means of securing cross-pollination,

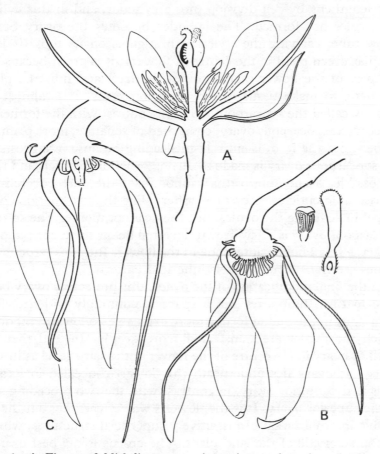

Figure 65. A, Flower of *Michelia montana*, in section, to show the reduction of the magnolia flower to three carpels (two shown) and few stamens; × 2. B, Flower of *Polyalthia purpurea* (Annonaceae) in section to show the magnolia flower with shortened axis, stamens and carpels, working as a trap to retain pollinating beetles in the cavity formed above the bulging inner petals; × 1½. C, Flower of *Mezzettia* in section to show how the annonaceous flower reduces to six stamens (three shown) and one carpel; × 7.

or outbreeding, rather than self-pollination or inbreeding. The little structure is a reproductive and discriminating antenna.

The next modification of the ovary is that which does away almost completely with the carpels (or megasporophylls) except for their active antennae or styles. The carpels are started at the stem-apex in the bud sufficiently to make the stigmas with their styles, but no more. Beneath these structures, set in a small ring round the stem-apex, a new growth of the stem itself occurs in the form of a ring that lengthens by cell-division into a cylinder, and in this cylinder the ovules are formed. The cylinder becomes the ovary box, or ovary tube, carrying the styles with stigmas on its tops. It is the familiar green pistil in the centre of flowers; it is green because it is the part of the flower that becomes the fruit and must be photosynthetic to make tissue for the growing seeds. It is almost universally called the syncarpous ovary in contrast with the former and antecedent apocarpous ovary, composed of separate, more primitive, carpels, but the term is entirely misleading because it suggests that the syncarpous ovary is made by fusing or joining carpels and this is impossible. Plant outgrowths cannot fuse: they are covered with indissoluble cuticles: they may adhere, like the two margins of the carpel in covering the ovules, but they cannot merge. The so-called syncarpous ovary is a new growth inserted below the carpels and it is, in fact, a small hollow internode in the flower. Into this structure the ovules are transferred in their inherited pattern.

In the final modification of the ovary, this projecting ovary box or tube in turn disappears and a space is cunningly widened at the stem-apex in the bud to form a more extensive hollow internode into which the ovules are transferred more deeply. Instead then of a pistil or ovary in the centre of the flower at the same level as its other parts, the ovary lies underneath the flower, concealed in its stalk. This is the inferior ovary, in contrast with the two preceding states of the superior ovary. It is the flower's way of carrying out the protective internalization of primitively superficial structures, which is so characteristic of the land plant. The crocus is the best example, for its ovary is not just below the flower but is underground, and removed by several inches of style from the receptive stigma.

As stout stems bear stout cones, so stout twigs should bear stout flowers. Primitive, pachycaul flowering plants should have stout flowers at the ends of these twigs. In fact this is very rarely so, and the only satisfactory examples are the flowers of the *Magnolia* family.

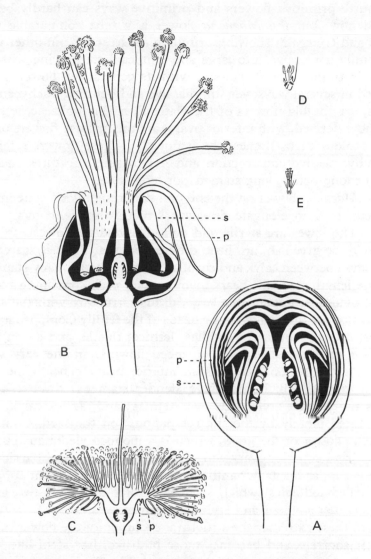

Figure 66. Sections of flowers to show the webbing of carpels into the syncarpous ovary and the sinking of the ovules into the flower stem to make the inferior ovary; *s*, sepals; *p*, petals. A, Flower-bud of *Dillenia*, in section, the carpels united but the styles free; note also how the parts of the flower fit into the spaces formed by the growth of the bud; × 3. B, Durian flower (*Durio*), with syncarpous ovary and united style, the ephemeral parts of this nocturnal flower about to fall; × 1½. C, Flower of *Eugenia* (Myrtaceae), in section, with inferior ovary; × 2. D, Grass flower, and E, flower of the Compositae to show in comparison their neotenic minuteness and simplicity; × 2.

Of course primitive flowers and primitive ways can hardly be expected now, but the *Magnolia* flower is a relic comparable with *Cycas* and *Ginkgo* in showing primitive construction; in other ways the family has grown into large and leptocaul trees vying with the tallest in tropical forest, yet still with its generalized flower. More derived in several ways, yet still massive and befitting pachycaulous plants, are the big flowers of the water-lilies (Nymphaeaceae), and still more derived, with inferior ovary, are the massive flowers of the pachycaulous cacti. Bigness, be it noted, in these flowers refers to generally massive construction and not to single features such as large or long petals, long stamens, or long style.

The *Magnolia* flower on the end of a twig resembles a cone in two respects: it is an elongate die-away structure set with many bud scales. The lower are sterile and petal-like and the lowest, though they may be greenish, are little different so that there is scarcely a distinction between calyx and corolla. Then come the many stamens, and the lengthened axis bears many carpels each containing two to twelve ovules. Compare this large, diffuse structure with the small slender flowers that make up the heads of the family Compositae (the daisies, asters, sunflowers, dahlias, lettuce, thistle, and so on), for they are an extreme of very advanced flowers. In the very short flower stalk is set one ovule as an inferior ovary beneath the very slender corolla, which bears five slender stamens; through them passes the slender style with two stigmas; and, as for calyx, it is represented merely by the silk (or pappus) on the seed. Contrast, then, the flower of the grass, which is a monocotyledonous perfection. It has a superior ovary with one ovule, two styles, three slender stamens and, as for its perianth, there are merely two slight appendages, called lodicules, which swell and force apart the two green bracts so that stamens and style may project.

From these extremes the structural evolution of the flower can be read. It shortens and becomes more bud-like, less stem-like. The ovary shortens to one ring of carpels and this allows the ovary tube to develop as the syncarpous ovary and, then, the ovules to be borne in the stalk of the flower as its inferior ovary. The flower becomes smaller. The number of its parts become less until they are represented by single rings of five, four, three, two, or, even one petal, stamen, or carpel. This allows the petals to be carried on a corolla tube, and the stamens on a staminal tube, similar to the ovary tube; it begets the telescopic flower with slender parts supported by special internodes that make the telescopic tubing. The number of ovules in the carpel reduces to one so that, just as the ovule (megasporangium) reduced to one megaspore and the embryo sac to one female nucleus as success became more certain, the flower reduces to one seed. Finally the sexes may separate and male and female flowers are

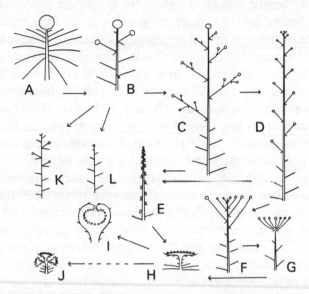

Figure 67. Evolution of the inflorescence. A, Massive terminal flower on a pachycaul. B, Less massive and with lateral (axillary) flowers. C, Panicle, and D, the raceme, as distinct inflorescences with the flowers in the axils of bracts. E, Spike or catkin, formed by the suppression of flower stalks. F, Corymb, as a raceme with the flowers brought to one level by lengthening the stalks, and leading, by loss of stem internodes, to G, the umbel. H, Head of flowers (capitulum), formed by condensing the spike or umbel and turning the inflorescence into a second order flower. I, Fig, formed by inverting the head of flowers. J, Head of heads (compound capitulum), formed by repeating stages B–G with the head of flowers in place of the one flower. K, Solitary axillary flowers, and L, solitary terminal flower, produced by the leptocaul habit of B.

produced to be borne, even, on separate male and female plants. So sexuality, referred from gametes to spores, is referred to flowers and then to the whole sporophyte, which means that the zygote from which it arises must be sexed. Sex, in fact, becomes thoroughly absorbed into the reproductive mechanism and is part of the chromosomal inheritance. It is this that makes the genetics of sex in the flowering plant comparable with that of the animal, though far from identical. The old asexual state of the sporophyte completely disappears.

As the flower becomes smaller so the inhibiting effect of the flower-bud lessens. The number of flowers developing from a stem or twig increases. Flowers appear from the buds in leaf axils as well as from the terminal bud. Then these flowering leaves become reduced to bracts and the leafy twig terminates in a branching inflorescence of flowers, no longer massive as that of *Magnolia*, though they may be showy with large petals. The inflorescences compact, through failure of internodal growth, into heads of little flowers, as in Compositae, and heads of flowers are set in branched inflorescences that may compact into heads of heads. Alternatively, the flowers compact on to a central stalk to form a spike, as in the palms, and the spike shortens till it is swathed by one large coloured bract and the whole inflorescence resembles a single large flower, as in the aroids. Yet further, in small plants, the branching inflorescence reduces to fewer flowers until a single one terminates the inflorescence stalk, as in tulips, daffodils, and some irises, and gives the impression of a primitive terminal position comparable with that of *Magnolia*. The history of the flower and inflorescence is a series of revolutions becoming simpler and simpler in construction, yet each revolution begins with vestigial inheritance from the preceding and by this means the sequence can be read. No longer is it thought that simple flowers with one stamen or one carpel are primitive. The single stamen and the single ovary with one ovule in the minute flowers of the most minute duckweed (*Wolffia*) are not the beginnings of the flower but the ending of the spike-like inflorescence of the aroids, just as its tiny leafless and rootless body is an aroid reduced to a minute sub-seedling existence. The physiological keys to flower evolution are the transference of function from one part to another of flower and inflorescence, neotenic substitution, and the varied effect of growth substances on the parts developing in the bud.

The flower brought a complexity into seed plants which they exploited to the full. Excepting the fructifications of fungi, no other plant structure has led to such enormous variety. It is the basis of the classification of flowering plants by which they are assembled into evolutionary alliances, and so clear are their many peculiarities that this classification was pretty well established by 1824, when De Candolle's great monograph of the flowering plants of the world was commenced, before there was any profound knowledge of the flower: this vast undertaking has not, however, been completed and we have no reliable inventory of the world's flora. The natural history of flowers, whether detailed in floral structure or floral working, has been the subject of many books, but it needs yet orientation from the tropical standpoint. It is, too, a strange reflection that the flower, which is the briefest bud, has dominated the human approach to botany. The flower must be seen in fuller perspective.

As mentioned on p. 152, flowering plants cannot be classified by vegetative characters. The evolution of flower and inflorescence must have been worked out before there was any considerable move to introduce the modern flowering forest and all its derivative growths. This is the lost, pachycaulous era which is the abominable mystery, too putrescible probably for fossilization. I am sanguine enough to feel confident that, with progress in plant chemistry and attention to these odd plants that have been neglected, we shall re-create the lost world: already plant breeding resuscitates pachy-caulous features, as in cauliflowers, in the belief that they are new.

The falling of pollen from stamens on to stigmas seems a trifling matter. The result is, however, self-pollination and inbreeding, against which there is as much prejudice in plants as in animals. In many flowers self-pollination occurs only in the absence of cross-pollination. As many, if not more, flowers guard against self-pollina-tion by devices such as self-sterility, the shedding of pollen before or after the stigma of the flower is receptive, the placing of anthers where pollen cannot fall on to the stigma, or the separation of the sexes into unisexual flowers. Flowers are full of devices, and a de-vice, in botany, means not something purposive but a construction with an effect that has been selected in course of time as a fit means of survival in the struggle for existence. Thus flowers have devices to take pollen away from the narrow gap between stamen and stigma, to carry pollen from one flower to another, and, preferably, from one plant to another, to secure outbreeding. For this purpose wind, animals, and, with some aquatic plants, water, are the agents: it would require genius in human affairs to make so much out of so little, to develop this brief passage of a few millimetres into a postal service of pollen grains between plants, to risk in such extravagance the microscopic cells that perpetuate the race, and withal to succeed. The plant has played upon animal curiosity and produced one of the best examples that distinguish the living from the lifeless; against the principle of least action, which is measureless decay, protoplasm is organized.

Since the spores of ferns and the pollen of conifers are conveyed by the wind, it might be supposed that first flowering plants were wind-pollinated. This cannot be proved, but what is certain is that all living wind-pollinated flowers reveal in their structure and that of the fruits that they form, and in their systematic alliance with others, that they have been derived from the insect-pollinated. If ever it was needed to show that facile assumptions about plants were erroneous, here is the example [73].

Insects visit flowers, conveying pollen. Wind blows it out of the flowers indiscriminately. Insects require some warmth for their flight and some recompense for their visits; wind is a shifting, if ever-ready, agent. The stigma of the wind-pollinated flower will catch few grains, but many will be rubbed on that of the insect-pollinated. Clearly there must be rules for each method. Wind-pollination needs small, light, smooth, dry, easily separable pollen-grains readily shaken from the stamen; insect-pollination needs adherent pollen secured economically where the insect will touch it. Stigmas of wind-pollinated flowers, of which the maize is an example, must be long or feathery to catch the grains; those of insect-flowers are rounded and compact so as not to obstruct the insect and not to be broken by it. The insect-flower, receiving pollen in quantity, can have many ovules for fertilization. The wind-flower, gathering a few grains, cannot afford to waste ovules, and it will tend to the final condition of one plumose stigma to one ovule. Far more pollen will be wasted by wind-flowers, which will favour the unisexual state with many small male flowers to overcome the wastage of pollen, and few small female flowers to receive; if the flowers were bisexual, there might be too many ovaries compared with pollen output and, if the flowers were large, there would be great wastage in unpollinated or under-pollinated flowers. Attractive petals, scent, and nectar are meaningless to wind-pollination but are the essence of insect-flowers. Thus, as wind is no speciator, wind-flowers tend to be those that have become small, simple, and inconspicuous, and form one-seeded fruits. Insect-flowers tend to conspicuousness, richness, variety, and many-seeded fruits, corresponding with the predilection of their visitors.

Oak, beech, hornbeam, chestnut, hazel, alder, birch, poplar, and walnut are the typical wind-pollinated trees of the north temperate region. They have many dangling, mobile, male catkins, out of which the pollen is shaken, and few small rigid spikes of few female flowers. These inflorescences are cone-like but they are cones of flowers, not primitive cones of sporophylls, and therefore derived super-cones or 'second-order' cones. These trees are also deciduous. They flower in spring before the new leaves, when there are no pollinating insects

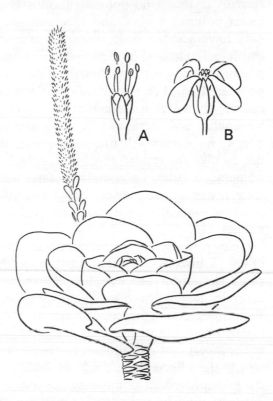

Figure 68. Kerguelen cabbage *Pringlea antiscorbutica*: × ¼. A, Flower of *Pringlea* with long stamens, and B, the flower of the stock *Matthiola*, with short stamens normal to their family (Cruciferae); × 2. (After Hooker, 1845, and Schulz, 1936.)

on the wing, and they shed their pollen before the foliage can obstruct its flight. Contrast the evergreen tropical forest teeming with insect life, where wind-pollinated plants are so few that they are hard to find. The bamboos, as tree grasses, the screw pines (*Pandanus*), and a few odd trees as *Casuarina*, and the tropical conifers, and perhaps, the allies of the walnut (as *Engelhardtia*), are examples. Among trees, wind-pollination is a device selected for the peculiarities of the winter season. Herbs suffer likewise. The salad burnet (*Poterium*) is a wind-pollinated herb of the insect-pollinated

Rosaceae; the meadow rue (*Thalictrum*) in the insect-pollinated Ranunculaceae tends to wind-pollination; the docks (*Rumex*) in Polygonaceae and *Artemisia* in Compositae are other examples. On the windswept Kerguelen Island in the south Indian Ocean there is the Kerguelen cabbage (*Pringlea*) of the insect-pollinated cruciferous family; it is wind-pollinated with projecting stamens and stigma, and the insects of the Island are flightless. The ash tree (*Fraxinus excelsa*) is the wind-pollinated, mostly northerly member of the insect-pollinated olive and lilac family (Oleaceae). Many tropical oaks have scented male flowers and nectar in the female and are insect-pollinated. The Spanish chestnut (*Castanea*) is both wind-pollinated and beetle-pollinated. The plantains (*Plantago*) are wind-pollinated and bee-pollinated. The witch-hazel family (Hamamelidaceae) shows all gradations from insect-pollinated bisexual flowers with petals and nectar to wind-pollinated, unisexual flowers without petals and nectar. All the wind-pollinated flowers have structural details, like the syncarpous ovary, inferior ovary, vestigial or supernumerary ovules, or vestigial petals, which prove their derivation from the more complex and manifold insect-flowers. So, among the most advanced monocotyledons, grasses and sedges are wind-pollinated.

Colour, shape, scent, and nectar are the attractions that the flower offers to insects. To these must be added, however, warmth and protection from the weather. By these means in almost endless diversity flowers have inveigled, detained, and despatched insects, sometimes even without reward [74]. So intricate is the relation, nevertheless, between the character of the flower and the character of the insect that it is difficult to decide which has been the cause and which the effect. Thousands of insect species are differently and daily employed in pollinating thousands of flower species; alternatively, these insects or their forebears have elicited as many flowers to satisfy them. The flowers may remain sterile without the insects, and many of the insects, such as bees, moths, and butterflies, depend on the flowers for their food. The subject of insect-pollination blossomed last century when naturalists could still know their plants and animals. Nowadays specialists have carried research on flower construction and on insect behaviour into unrelated depths, but they will be reunited as the pendulum of biology swings back to field-work. Botanists must learn to study flowers with the new knowledge of insect vision. A flower in polarized or ultraviolet light looks very different to the insect eye, which can thus see it, and the human eye, which cannot. Botanists and zoologists must transfer their researches to the tropics. Honey-bee flowers of Europe are very different from palm flowers, which appeal to the tropical bees. Nevertheless, the introduction to Knuth's great summary of the studies of last century still makes fascinating and productive reading: it provides, too, the systematic comparison that is so necessary for clearness in this example of parallel evolution [75].

Magnolia flowers are pollinated by beetles, which, as flower insects, are clumsy and indiscriminate. They bite and break. The large flowers of water-lilies are pollinated by beetles. Those of the giant *Victoria amazonica*, 30–40 cm. wide, are visited by cockchafers, which eat the stamens and then lay up in the flowers when they close in the morning until the ensuing evening [76]. The allied and ubiquitous tropical family, the Annonaceae (to which custard apple and soursop belong), has varied this theme into thousands of species of flower, largely employing small beetles which are often trapped overnight or overday. Hence grows the idea that primitive flowers, offering stamens as food, and perianth as protection, were first beetle-pollinated, and these creatures, as insects go, are not particularly advanced. Later came the smaller flowers, or those with narrow tubes, which supply nectar and fit not insect's bodies but their tongues, backed with greater perspicacity. Thus roses, poppies, anemones, rock roses, and other widely open flowers are in the first place beetle-flowers. Maybe the smell of pollen, or the smell of the opening flower which is always there, attracts; and it has been suggested that this was the primitive scent to the olfactory insect. Maybe the smell enticed beetles to mating and egg-laying, as happens, it is said, in the female cones of the African cycad *Encephelartos*.

From open scramble flowers there have been two main lines of specialization, and every specialization, be it noted, diminishes the variety of insect visitors and lessens the chances of pollen going astray: insects become specialized to the specialities of the flower and traffic between its kind. One line leads to small and numerous flowers presenting an array of little nectar cups, none of which is filled to satisfaction, and culminates in the heads and spikes of the umbellifers, composites, and palms. The other leads to inflorescences of relatively few large showy flowers befitting large insects that prefer a front entrance; as the single flower succeeds, so its eggs, as it were, are put into one basket.

Large flowers become bilaterally symmetrical or, as botany has it, zygomorphic. They develop an upper lip, which serves as a hood for the stamens and stigma, and a lower lip where the insect alights. With one entrance, the numerous stamens of radial symmetry, projecting in all directions, are uneconomical and the stamens reduce to two or one placed where their open anthers contact the insect's back

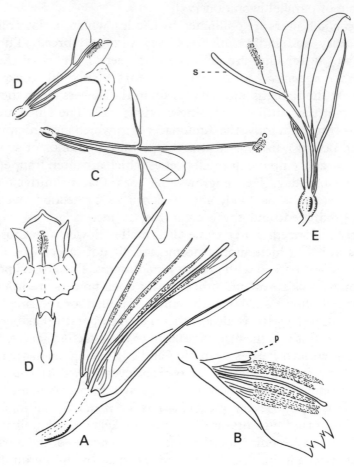

Figure 69. Specialization of the flowers of the banana alliance Scitamineae; natural size. A, Bird-pollinated traveller's palm *Ravenala*, with six long stamens. B, Bat-pollinated banana *Musa* with five short, thick stamens; p, free petal, the other two attached to the three-pointed calyx. C, Moth-pollinated ginger *Hedychium*, its one anther clasping the stigma. D, Butterfly-pollinated ginger *Curcuma*. E, Flower of *Canna*, with all the stamens turned into coloured petals but one bearing an anther; possibly bird-pollinated.

and the stigma is put where it will rub against it. Thus the large symmetrical flowers of lilies lead to the precise flowers of orchids, in which one stamen survives with its pollen agglomerated into one or two masses in certainty of success. Around this construction are ranged some ten thousand species of orchids, every one of which by colour, size, shape, scent, and time and season of opening attracts its own kind of insect and achieves its own postal service. When this is broken down and pollen is misdirected by artificial hybridization, all manner of hybrid orchids can be raised in horticulture; in nature, however, the specialization of the pollinator prevents promiscuity. The orchids are paralleled by another group of monocotyledons, the tropical gingers (Zingiberaceae), which have arrived at a similar conclusion by a different route with more vivid and delicate flowers. Among dicotyledons, the most wonderful array occurs in the caesalpinioid trees and climbers of the bean family (Leguminosae). Here belong *Cassia, Bauhinia, Amherstia*, and a hundred more genera of the tropical forests, displaying the zygomorphic specialization along several different lines to end in three, two, or one stamens.

Bees clutch and crawl; butterflies and moths hover and flutter. Stiff flowers suit both, but the bee enters beneath hidden stamens and stigma, whereas slender projecting stamens dust the butterflies and moths when they sip nectar with long tongues. Length of corolla tube and length of insect tongue are another interplay, long tubes excluding short tongues, so that naturalists have justifiably postulated very long-tongued lepidoptera when they have found flowers with tubes a foot long, as in some bignonias and orchids (as the Madagascan *Angraecum*). DDT is the great invention but, in exquisite sympathy, honeysuckle is the moth's midsummer-night's dream.

As if they were without prowess in these matters, to silly flies with short tongue and light body the temperate botanist attributes small, green, white, and yellowish flowers, like those of holly, maple, and spindle tree. In the tropics fly-flowers are the biggest of all creation. They have a lurid appeal, where big things please these little minds. To a dark hole in a fleshy composition, suffused meat-red to a decaying purple, or blotched on white, frilled, tailed, spiked, even set, it is said, with a glittering mirror, a stench invites the bluebottles, carrion flies, and dung flies. In the belly of the flower they buzz until some alteration in the softening of backwardly pointing hairs or other change in the corolla allows them, dusted with pollen, to escape. *Aristolochia* is a climber with such a flower curved in a tube. *Rafflesia* is the leafless parasite of vines in South-East Asia with the most massive flowers of all, the Sumatran *Rafflesia arnoldii* being 3 ft. in diameter, and these flowers squat like fungus on the ground. The aroids, however, have transformed a whole inflorescence into a compound flower with one large coloured bract, by which means thousands of species have again attracted as many flies and little beetles to their specific attention. One aquatic aroid (*Cryptocoryne*) projects this bract as a tube some 12–18 in. through the water to open at the surface, and down this shoot slip beetles to overnight in an underwater cabaret (figure 100).

The story is not complete. If beetles lead to better insect-pollination, what of birds? Are their flowers glorified insect-flowers or have primitive flower birds evolved with their class of flowers? There must be no prejudice because nothing is certain, if anything be known at all, of the early evolution of forest birds. Brilliant, strong, elastic, scentless, and supplied with copious watery nectar are the flowers of their preference. According to Pickens, the descending order of the bird's choice is red, pink, orange, blue, yellow, white, green, and maroon [77]. The human eye can appreciate bird vision where it fails to understand the insect's, which cannot blend. The birds are, mainly, the humming birds (Trochilidae) and honey-creepers (Coerebidae) of the Americas, the honey-eaters (Meliphagidae) and brush-tongued parrots (Trichoglossidae) of Australia, the spectacle birds (Zosteropidae), nectar birds and sun birds (Nectariniidae), and flower-peckers (Dicaeidae) of Africa and Asia, and the honey-creepers (Drepanididae) of Hawaii. The birds are rougher than insects, and delicate flower parts, such as the ovary, are more often concealed in the inferior position, or in some way separated from the nectar, as the stalked ovaries of the passion flowers [78].

Mistletoes (Loranthaceae) as bird-flowers can be understood; they hang from branches of tropical trees in several hundred species, their flowers with avian brilliance, and as the fluttering bird pecks at the bud it pops open and scatters over its head the pollen. Other families of plants present a problem. The large-flowered tropical trees of the Leguminosae are bird-pollinated; the smaller and the temperate ones are insect-pollinated, though even the tiny-flowered acacias may be bird-pollinated. Thus, besides the caesalpinioid flowers, such as *Caesalpinia*, *Delonix*, *Bauhinia*, *Brownea*, and *Amherstia*, the tropical coral trees (*Erythrina*) and many others with papilionaceous flowers are bird-pollinated. Because of the copious nectar which drips from its inverted inflorescences the American *Erythrina crista-galli* is called the 'cry-baby flower'. In East Java, the flowers of *Erythrina*, produced in the dry season, supply the main source of liquid for birds as large as crows. The big flowers of the silk-cotton trees (*Bombax*) and of *Hibiscus* are bird-pollinated; smaller and temperate flowers of the same alliance, as mallows and hollyhocks, are insect-pollinated. In his detailed study of Ericaceae (rhododendrons, heaths, vaccinium) and of the South African and Australian Proteaceae, Werth was forced to conclude that the large bird-pollinated flowers of the tropical and subtropical kinds were antecedent to the smaller and, mainly, temperate insect-pollinated [79]; in the cheerless summer of the Faeroe Islands a wingless thrips pollinates the heather. There is the same inevitable conclusion in the order Scitamineae of monocotyledons, which begins with the traveller's palm (*Ravenala*) and *Strelitzia* and ends with the insect-pollinated herbs. Temperate botany views mullein, snapdragon, foxglove, and penstemon (Scrophulariaceae) as insect-pollinated; tropical botany contributes as their antecedent the *Bignonia* family of large-flowered trees and climbers, most of which are bird-pollinated. So stand Capparidaceae to the insect-pollinated crucifers, Lobeliaceae to Campanulaceae, Verbenaceae to Labiatae, and *Fuchsia* to *Epilobium* (Onagraceae). Even the South American *Mutisia* and its allies, with rather large flowers, are the bird-pollinated Compositae. No serious student of flower-pollination can escape the great body of evidence which suggests that there has been an evolutionary sequence of pollination, from the Cretaceous period, by birds, then insects, and finally by wind or, more rarely in aquatic plants, by water. What we need is tropical natural history to understand from what tropical source the pollination methods of the temperate regions have been derived, and what may be the systematic and generic transition from one kind of pollination to another. It may be quite wrong to suppose that clover, mint, and Canterbury bell are primarily bee-flowers evolved by insects. The magnolia and water-lily may suit beetles, but that does not mean that they evolved them, any more than thrips the heather.

Bird-flowers grade also into nocturnal flowers pollinated by small bats. Round the night-blossoms of some cacti (*Cereus*) dart small vampires, hooking by their claws on to the flowers, thrusting their noses into the masses of stamens, taking the nectar and, incidentally, pollinating. The calabash tree (*Crescentia*) of America, the sausage tree (*Kigelia*) of Africa, and the midnight-horror (*Oroxylon*) of Asia are bat-pollinated bignoniaceous trees. The nocturnal baobab (*Adansonia*) of Africa, and the cotton trees (*Ceiba*) and durian trees (*Durio*) of Asia are bat-pollinated members of the *Bombax-Malva* set of families. Some nocturnal caesalpinioid flowers of America, as *Bauhinia megalandra* and *Eperua falcata*, are bat-pollinated; so may be the Asiatic members of the mimosoid *Parkia*, whereas the American with red and yellow colours are probably bird-pollinated. The papilionaceous *Mucuna* has bird-, bat-, and insect-pollinated flowers. The European purple loosestrife (*Lythrum*) is an insect-pollinated ally of the bat-pollinated tropical trees *Duabanga* and *Sonneratia* of Asia. Red-flowered wild bananas are said to be bird-pollinated, those with purple and brown flowers bat-pollinated.

Bat-flowers are also large, strong, as fitted to the bat's claws, and copiously suppled with watery nectar. They differ from the bird-flowers in opening at dusk, in their lurid yellow–green, red–brown, or purple–brown colours, in their wide mouth, allowing roomy entrance, and in the strong smell, which may be a foxy or fishy stink, a smell of sour milk, or a strange essence of cucumber. Often, too, these flowers are borne in hanging inflorescences, or they face downwards, suiting the bat's posture. They are bird-flowers adapted to bats, as the brightly coloured butterfly-flowers adapt themselves to moths [80]. But what of the history of this association? Bats eat fruits, sup nectar, and catch insects; thus they impinge on the life of plants. The answer is not in books, which are now too specialized, but waits the biologist who can appreciate both plants and animals in tropical life.

Chapter 12

The Fruiting Tree

THE EFFECT OF the pollen-tube on the flower is manifold. It causes the embryo, the endosperm, the seed coats, and the fruit to develop in their particular ways; it stimulates the stalks of the flower and inflorescence to thicken and supply the increased demand for food; it causes, also, the parts of the flower that have fulfilled their function, such as stamens, petals, and sepals, to wither and fall off. It is well known that orchid flowers, which will remain fresh for several days or weeks, soon wither after pollination. It is known, too, that the pollen of one species placed on the stigma of a related species may cause fruits and even seeds to form but without embryos, because the pollen has been unable to effect conjugation or, after conjugation, to produce a viable embryo. If the pollen-grains alighting on the stigma are too few to fertilize all the ovules, then fruits form that are variously undeveloped, waisted and notched where seeds do not form, as may happen in cucumbers and bean pods, or lobed and lopsided where seeds have formed, as in horsechestnut and bread-fruit trees. Some long fruits, however, are habitually constricted between the seeds, showing where the ovary wall has enlarged round the seed and failed to enlarge where there is no seed: such are the bean pods called in botany 'lomenta', of which the variety of cow pea (*Vigna*) called in Malay *kachang perut ayam* (chicken's-gut pea) is a good example.

The underlying cause of these manifestations is now considered to be the production of growth substances in the ovary or in the ovule, perhaps mainly from the young endosperm, as a result of the growth of the pollen-tube, and the spread of these chemicals to other parts of the flower, causing them to wither and fall off or to grow in their particular way. Thus if the growth substance auxin, in very dilute solution, is sprayed on to tomato flowers it is absorbed and causes them to form fruits that are normal except for the absence of seeds on embryos. Other products of fruit culture have this faculty in them and, as some oranges and the majority of cultivated figs, they normally produce fruit without pollination. Here is the beginning of chemical understanding of fruit growth, but it does not explain why the various parts of the flower behave in their particular ways after pollination, for these are the very complicated effects of inheritance modified in the course of evolution.

The first effect of pollination on the ovary is to restart its growth. The ovary extends to the full length of the fruit and then swells with the slower growth of the seeds. These have also their characteristic sizes. They will vary somewhat but within certain limits; whether or not they fill the interior of the fruit, they have a specific size. Yet, there are exceptions, such as horsechestnuts (*Aesculus*) and peanuts (*Arachis*), the seeds of which differ greatly in size according to the number (three, two, or one) developed in the fruit. Likewise, some normally one-seeded fruit, as acorns, avocado, palm fruits, or grass 'seed', will show considerable difference in the size of the true seed, which depends on the size of the fruit. Such are overgrown seeds that never mature with a proper seed coat but employ the fruit wall instead, and fill the fruit to maximum capacity; they resemble benign cancers [81].

Lastly the fruit ripens. Its cells become old and moribund and, according as they have grown and diversified into tissues, so that fruit has its mature character distinctive of the kind of flowering plant that it represents. The ovary of the flower is, clearly, a structure halted temporarily for pollination, and the fruit is the final expression after pollination of this die-away reproductive bud. By their fruits they shall be known, yet the study and explanation of fruits is the chapter in plant life most in need of improvement.

Seed cones of conifers are comparable. They halt at the receptive stage when the scales diverge and admit the pollen that is blown on to the exposed ovules. The cones then enlarge and, on ripening, dry up and open the scales for the detached and mature seeds to slip out. Very different are the seed cones of cycads. They do not halt. They grow continuously to their full size, when the ovules are as large and as fully constructed as they ever will be. Then they are pollinated and the slow fertilization of the egg-cells within generally occurs when the seed has fallen to the ground. *Ginkgo* seeds have the same manner of pollination at full size, to be followed by fertilization of the egg

27. The dagger-tree *Pajanelia* (Bignoniaceae) with improved pachycaul habit, but very long pinnate leaves; at Baling, Kedah.

28. The wild breadfruit ally *Artocarpus elasticus* with large simple leaves, in Negri Sembilan; see 16 (b).

29. A giant dipterocarp *Anisoptera*, over 200 ft. high, with a strangling fig on its trunk; a relic of primary forest in Negri Sembilan: telephone pole for scale.

30. The pachycaul Composite tree *Espeletia* in the Columbian Andes.

31. Pachycaul curiosities: (a) the melon tree *Dendrosicyos* of Socotra;
(b) the Apocynaceous tree *Adenium* of Socotra; (c) the
Apocynaceous tree *Pachypodium* of East Africa, cactus-like without
internodes and with small leaves only on the top of the stem.

32. A long-nosed bat *Leptonycteris* pollinating the flowers of the century-plant *Agave*. New Mexico.

b

33. (a) the thorny swamp-palm *Zalacca*, in Singapore; (b) its unopened
leaf with the thorns laid flat against it in a familiar
basket-pattern of the East.

34. A giant aroid *Colocasia* in the forest of Mt. Kinabalu, North Borneo.

35. (a) *Casuarina* forest advancing over the coastal dunes of Pahang into the preceding zone of herbaceous plants; (b) primary forest at high tide level on the Johore coast. *Scaevola* (bush, left foreground), *Cycas* (in recess, central), *Calophyllum* (tree, right foreground), *Hernandia* (above *Scaevola*), *Terminalia* (above *Cycas*).

a

b

a

b

36. (a) the big lily *Crinum* in the mangrove forest of Johore; (b) the sago-palm *Metroxylon* in Negri Sambilan; (c) the swamp-palm *Nipa* by a tidal river of Johore.

37. Trees re-invading deforested and burnt land between the forested rivers; Araguari drainage basin, Amapá, Brazil.

38. Amazon forest delineating river-changes with lines of trees; the muddy Rio Madeira, and the dark peat-water lakes being invaded with floating vegetation from the edges.

39 (a) the forest water-lily *Barclaya* with heart-shaped leaves (right foreground) growing with the ginger *Plagiostachys* (in fruit); Johore; (b) the airy crown of the dipterocarp tree *Dryobalanops*; Malaya.

a

b

40. Rivers with fringing (gallery) forest, intervening land deforested and burnt off to semi-desert; Goiaz, Brazil.

41. (a) the flowers of the durian, *Durio zibethinus*, hanging from the branches; (b) the durian fruit; (c) the fruits of *Durio testudinarum*; (d) the flowers borne at the base of the trunk. Sarawak.

after dispersal. In fact there is no distinction between ovule and seed: or better, one should say that the seed is pollinated and there is no ovule.

Hence arises the theory, disputed time and again but the only competent explanation, which sees in the carpels of the flower very immature seed leaves (female sporophylls) working prematurely for pollination while still folded up and protected in the centre of the flower-bud, though the rest of the bud has opened: and then, only after the stimulus of pollination, do the carpels grow to their full size, unfold, and display the seeds for dispersal. By this means unpollinated flowers, lightly made, do not waste the mass of material needed for a fruit that would be without embryos; such unsuccessful flowers fall off inexpensively and the resources of the plant can be devoted to those which are pollinated. According to this theory the central ovary, which is the last part of the flower to be formed, has delayed until pollination the production of the growth substances that continue its development. If *Cycas* and *Ginkgo*, so archaic in their retention of sperms, explain the beginning of the seed, then the orchids, with one of the most advanced of flowers, explain the end of floral evolution. In many orchids the ovary is so immature at the time of pollination that it has not even formed ovules: substance is not wasted even on them unless pollination is secured. The same delay happens in various other plants, such as the tropical 'cherry' *Muntingia* of the Tiliaceae. In other words, the flower is the reproductive bud that opens precociously for pollination, after which its immature centre grows to its full extent and converts tentative and immature seeds (ovules) into mature seeds. What was continuous growth to the seed becomes halved into a flowering growth for pollination and a fruiting growth for dispersal. The flower is another example of neoteny or the working of a structure before it is fully developed. Thus we speak of flower-buds, but for the ancestors of flowering plants we should speak of fruit-buds.

As sepals recurl, petals unfold, and stamens lengthen to open, so after pollination carpels lengthen, swell, and unfold to release the seeds; until ready, the developing seeds are protected in the unopened ovary. In the ovule there are buried the relics of the female spore and female gametophyte; in the ovary there is buried the ovule until, after pollination, a trace of normal leaf behaviour in the fruit sets it free, or externalizes it to start the new generation. In the tropical forests of the Old World there are a few trees in which this

14

fruit device is imperfect. The carpels, five in a flower, quickly become green and baggy after pollination and they soon open to expose along their edges the immature seeds, which continue their growth, thus displayed, for several weeks. The trees belong to a small alliance in the Sterculiaceae, namely *Firmiana*, *Pterocymbium*, and *Scaphium*. Were they extinct and their open carpels with immature seeds known only from fossils botany would take notice, but alas, living and indeed lofty, their fruits are unheeded as museum curios!

The seeds of these three kinds of tree are never detached from the fruiting carpels. They dry up, separate from the fruit stalk, and serve as a parachute which blows away and flutters in spirals through

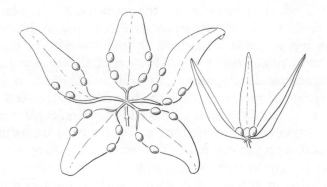

Figure 70. Immature fruit of *Firmiana* (left) with all five green carpels or pods opened to expose the developing seeds; × ½. Right, similar fruit of *Scaphium* with one seed developing in each pod; × ¼.

the forest to reach the ground where the seeds germinate. If, as in *Firmiana*, several seeds are carried on the parachute, they sprout together and compete in their growth until one survives. If, as in the two other genera, only one seed is carried, this wastage of seedlings is avoided; and a survival value in the struggle for existence can be ascribed to those trees, which, having undetachable seeds, tend in their fruits to restrict the number of the seeding ovules to one.

These trees introduce four positive features into the study of fruits where their significance is all-pervasive, namely dehiscence (fruit opening), undetachable seeds, wind-dispersal, and one-seeded fruits. A well-defined starting-point is necessary because fruits are bewildering in their diversity. Many families have their particular kind of fruit, which will depend, of course, on the kind of ovary from which it formed, whether apocarpous, syncarpous, superior, or inferior. In others, as the Rosaceae and a great many families of tropical trees, the fruits are characteristic of their genera or of groups of genera. It was Linnaeus's intention that genera should be defined on fruits, and it may be right to say that there are as many sorts of fruit as there are major groupings of genera. This means thousands of kinds and, indeed, several hundred special names have already been proposed; the result is another specialist language in the babel of science, through which the student must break.

Dehiscent fruits open and let the seeds out for dispersal. The common example is the dry capsule, which is formed from a syncarpous ovary. Fruits that do not open but contain the seeds are called indehiscent. There is plenty of evidence that they are derived. Dehiscent fruits always split along special lines, called sutures, which signify structural features in the ovary corresponding with carpel position. The sutures persist in many indehiscent fruits but they are operated now, not by the force of the active fruit wall itself, but by the swelling of the seed on germination. The outer velvet coat of the almond, which is a dry sort of plum, shrinks and splits along a suture normal for a single carpel and thereby exposes the stone or inner woody wall of the fruit: when the seed germinates it splits the stone clearly along a corresponding suture; plums and cherries differ in that the outer part of the fruit is pulpy and no longer splits. Their genus, *Prunus*, must have come from an ancestor with a carpel that split open to expose the seed just as in these sterculiaceous genera. In the eastern tropics, rambutan, and pulasan (*Nephelium*) correspond but a thin fruit wall covers in them a pulpy seed: this skin, on drying, very tardily splits along a suture which in other genera of the family (Sapindaceae) is actively opened to expose the seed. The walnut is the fruit of an inferior syncarpous ovary; its green rind often splits rather irregularly into two or four strips which correspond nevertheless with sutures of the stone or inner fruit wall, and these stone sutures are burst by the germinating seed; the vestiges of dehiscence are clearer in the American pecan (*Carya*). Wild bananas are botanically berries because their fruits contain many seeds within an edible, fleshy fruit wall. A berry is an indehiscent fruit derived from a superior or inferior ovary. A wild banana (*Musa schizocarpa*) of New Guinea and New Britain splits open from the top downwards into three lobes corresponding with the three-carpel form of the ovary and exposes the seeds, thereby showing how this kind of berry is derived not from a dry capsule, but from a fleshy capsule. In other families, such as the grasses and composites, where the fruit is uniformly one-seeded and indehiscent, it is argued by analogy that the fruit is derived from the dehiscent. Palm fruits, though never dehiscent apparently even in a vestigial way, may have nevertheless three or more seeds, which indicate a previous many-seeded and capcular ancestor.

The last point adds to the conclusion that one-seeded fruits are derived from many-seeded. There are numerous cases, such as anemones and clematis, where the carpel contains several ovules but one only forms a seed. The acorn, derived from an inferior ovary with three or four cavities and two ovules in each, is nevertheless a one-seeded fruit. Almonds may have twin seeds, as evidence that two ovules may occasionally develop in fruit. Some one-seeded

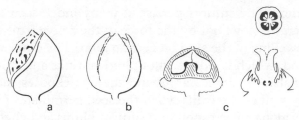

Figure 71. Tardily dehiscent rind of (a) the almond (*Prunus*) and (b) the pecan (*Carya*); c, perfected nut of the oak (*Quercus*), indehiscent, woody, and one-seeded, acting as a single seed but derived from an inferior ovary with six to eight ovules; × ½. Flower of *Quercus*, × 10; ovary in section, × 20.

fruits dehisce, such as nutmegs, but one-seededness has led in most cases to indehiscent fruits as drupes (fleshy, with a stone, as in plums), or nuts (which are dry), or achenes, which is a general term for small, dry, one-seeded indehiscent fruits.

The one-seeded fruit is indeed a second-order seed or super-seed, for it is a seed within a fruit coat. It represents another of the revolutions and involutions of plant evolution: instead of a flower with one large fruit containing many seeds, an inflorescence of many flowers produces many small one-seeded fruits; the total output of seeds may be the same, but the inflorescence offers more opportunity for pollination, in so far as the flowers open at different times, and suffer less loss if some flowers are not pollinated; it is the analogy of spreading the eggs over so many baskets that, at length, there is only one egg for each. Botanically it is the application to flower-fruits of the principle of singleness, begun with the egg-cells in their reduction to one in the gametangium, carried on to the single female spore in the female sporangium, the single egg in the female prothallus, the single female sporangium on the female sporophyll (carpel), and either the single carpel in the flower or the single ovule in the syncarpous ovary. Always in the course of reproduction in the plant a scene with many actors becomes monopolized; from manifold necessity comes singleness of purpose.

Is this all, or how may we know more? If there were very few plants, such as one kind of seaweed, one moss, one fern, one cycad, or one flowering plant, we should be stumped, and there would be no theory of evolution. As it is, there are many and very many, which, because they exist, are different states of evolution at these different moss, fern, or flowering plant levels of organization; and, if we are to understand them, we must compare them. We must explore the world's resources if we are to understand plant life. What kinds of fruit have the allies of *Firmiana*?

We come to *Sterculia* of the same family. Its flower has also five free carpels but in fruit they form a red star of five rather fleshy pods, which open to contrast many shining black seeds against the red of fruit and the green of foliage. The seeds are not detached but in a very short time after dawn in the forest they are pecked off and swallowed by birds, which grind off the thin pulp around them and void the hard part of the seed unharmed. The star recalls the rosette of seed leaves of the genus *Cycas* itself, which has no compact female cone, but the red fleshy seeds of *Cycas* sit on the greenish female sporophylls; in *Sterculia* black, thinly fleshy seeds sit on red female sporophylls against green foliage in heightened contrast and fuller evolution. The fruiting structure of *Cycas* was exposed during its development for several months: that of *Sterculia* works in as many hours.

Is this all? By no means, for there are over two hundred species of *Sterculia* from which we may enquire! In a few of them the seed is more elaborate. A thick layer of pulp arises from the base of the seed and covers it with another and red, orange, or yellow, jacket, which strengthens the whole colour effect. This jacket is comparable with the pink pulp of the yewberry (*Taxus*) or the orange pulp of the seed of the spindle tree (*Euonymus*). It is the part of the fruit that animals eat, but it is a part which seems always to be developed after pollination. It is called the aril and arillate fruits of this kind are generally so

213

rare that I have never seen a living one of *Sterculia* and I have been unable to obtain a colour photograph. If they were common and modern, like nuts, drupes, and achenes, there would be no need to illustrate them. Most species of *Sterculia* have no arils; a few have very minute arils at the base of the seed; still fewer have the seeds more or less covered by the edible aril. Merely one or two other genera of the family have arillate fruits: the great majority, at least a hundred, which comprise several thousand species, have no aril at all, but dry fruits with dry seeds, or indehiscent fruits, which are, as we have seen, derived. This aril is generally regarded as a modern ornament to the seed. It does not seem so in this big family of tropical plants. It seems from its rarity to be something primitive from which the other kinds of fruit have been derived. And, when we turn for comparison to other families of flowering plants, we find in general the same state of affairs; that is the rare arillate fruit and many derived fruits without arils, the evolution of which from the arillate kind may easily be followed among their genera.

For instance, in the bean family Leguminosae there are comparatively very few genera of trees with arillate pods, black seeds, and red or yellow arils. In the *Mimosa* group of forty genera, there are only *Acacia* and a close ally *Pithecellobium*, and most species of these genera have pods without arils, which lead to indehiscent pods. In the *Caesalpinia* group of one hundred and twenty genera there are five or six arillate genera, in most of which also there are dry pods without arils. In the bean group itself (Papilionaceae), with three hundred genera, only *Arillaria* has the typical fleshy pod and arillate seed, whereas most other genera have more or less vestigial arils making the tumid attachment of the bean seed. *Arillaria* is known to us by one kind of rare tree of Burma. In the orange family Rutaceae merely a few rare Sino-Himalayan species of *Zanthoxylum* have black seeds and red arils, though many genera have inedible dry capsules with small black seeds and no arils. Among the very numerous members of the tropical trees (Annonaceae) it seems that only some species of *Xylopia* have arillate pods: the majority are indehiscent and drupe-like without aril. In the big group of Malvales, covering several hundred genera and thousands of species, there are merely a few genera of tropical trees with the arillate fruit and, again, by no means all of their species. These are a few instances of this rare fruit.

Now a rarity like this among so many large groups of plants is not a sign of promotion. Most edible fruits with animal-dispersed seeds are indehiscent berries and drupes, derived, advanced, closed-up, and of a common occurrence to mark their progress. For example, there are plums, apples, raspberries, gooseberries, currants, grapes, jujubes, oranges, pawpaws, melons, passion fruits, mangosteens,

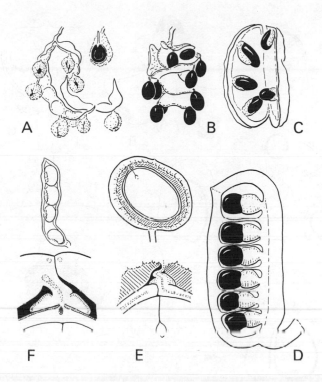

Figure 72. Story of the bean, revealed by the fruits of tropical leguminous trees; pods in section, × ½. A, More primitive pod with arillate seeds, *Pithecellobium dulce*. B, Modification with thinly fleshy seeds and no arils, *P. ellipticum*. C, Woody pod of *Pahudia javanica* with the black seeds partly covered by the aril. D, Very woody pod of *P. cochinchinensis*, the black seeds without aril but with a red fleshy stalk. E, Indehiscent plum pod of *Detarium senegalense* with a vestige of the aril at the end of the seed stalk within the hard stone; enlargement, × 2. F, Dehiscent pod of the bean *Mucuna* with small aril; enlargement, × 3.

lichis, mangoes, olives, sapodillas, persimmons, tomatoes, figs, mulberries, bananas, and palm fruits. They are so commonplace that we are apt to think that they could have arisen in any sort of way by making a fruit fleshy, but, just as in the study of flowers and pollination we learnt that facile assumption would not fit the facts, so with the fruits we must turn to the facts of nature and rediscover the truth, as I have elaborated it in the durian theory [82].

Briefly this theory states that the primitive fruits of flowering plants were dehiscent arillate carpels or, in the syncarpous ovary,

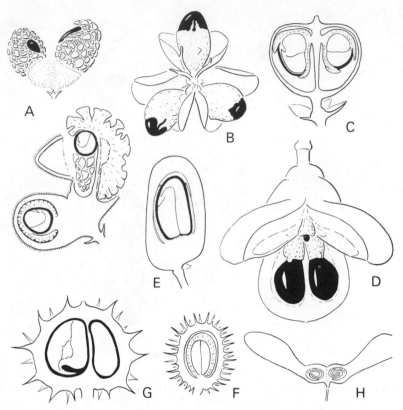

Figure 73. Explanation of the fruits of the horsechestnut and maple by those of tropical trees in their alliance of Sapindales; the black or dark-brown seed coat in black or as a heavy line. A, Fruits of *Alectryon* (New Guinea) with the top split off by the expanding crimson and tesselated aril; natural size (section × 1½). B, *Guioa* with the seeds more or less covered by the orange aril; × 2. C, *Harpullia* (New Guinea) with the aril restricted and the sides of the seed becoming fleshy; × 1½. D, Malayan *Trigonachras* with fleshy base to the black seeds and no true aril; natural size. E, Malaysian *Pometia*, the indehiscent plum-like fruit, orange coloured, with thin and concealed aril (dotted); natural size. F, Malayan rambutan *Nephelium* with spiny, indehiscent fruit and no aril, but pulpy seed coat; × ½. G, Horsechestnut *Aesculus*, spiny, without aril, inedible; natural size. H, Maple *Acer* with indehiscent fruit dividing into two winged nuts (samaras), without aril; natural size.

capsules. These fruits had bright red or orange fleshy walls and black seeds covered with red or orange edible arils, and the seeds were not detached until pulled off by animals. From such arillate fruits all other kinds have been derived by loss of the aril, by drying up into the rattling pod or capsule with its dry seeds, by indehiscence leading to the general succulence of berries, by increasing lignification leading to stone fruits and nuts, and by decreasing size leading to small

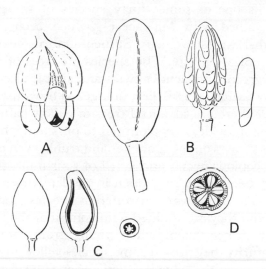

Figure 74 Modifications of the arillate fruit in the mahogany family Meliaceae. A, Fleshy arillate capsule of *Dysoxylon*, the black seeds covered by the red aril; × ¼. B, Dry woody capsule of the mahogany *Swietenia*, the outer wall splitting into five parts and revealing the winged seeds about to take off; × ¼. C, Indehiscent plum of *Melia excelsa* with very thin stone and one large bag-like seed; × ⅔; the ovary of the flower in section, × 3. D, Indehiscent plum of the Persian lilac *Melia azedarach* with very woody stone and five seeds; × ⅔.

editions of all these kinds. The modifications have occurred in most of the big families of flowering plants and have led to parallel products in all, so that the dry capsules, berries, drupes, and nuts of one, say, the orange family, are similar to, but in all details of construction different from, those of another, as the rose family. Walnut, chestnut, cashew nut, peanut, illipe nut, kenari nut, and coconut are parallel products evolved independently in their separate families, as are the tiny one-seeded fruits of buttercups, strawberry, parsley, dandelion, nettle, grass, and sedge. It is in their arillate fruit that the families, which still retain them, converge and point back to the beginning when pollination and seed dispersal by animals were one complex as they still are to some extent in the cycads. Here is the last great adventure in the lives of green plants whereby they have come to spread by different means over the land, and have worked out according to their family character their individual answers. Instead of a cabinet study in curiosity, there is the living science, or natural history, of fruit and seed with evolutionary direction.

The durian is one of some thirty species of the genus *Durio* (Bombacaceae, allied with Malvaceae), which occur in the rain forests of South-East Asia, chiefly Borneo. On this genus centres a great deal of tropical biology. The species are trees of moderate to great size, evergreen with slender twigs and simple alternate leaves, the undersides of which are silvery, orange, or brownish from small scales. They are therefore advanced trees of willow habit. The massive flowers, 2–4 in. wide, vary from red to pink and white and they are borne in most cases on the branches and trunk, even near its base. They produce copious nectar and smell of sour milk; they are pollinated both by bees, when they open in the late afternoon, and by bats, at night. The fruits vary in different species from 2 to 8 in. long and wide and they are studded with conical or thorn-like spikes that effectively protect them when young from interference by animals. At maturity they turn in some species yellow to red and split into five parts, along the edges of which hang the dark-brown or black seeds with red arils. Other species may have short arils or none at all, in which case the fruits turn brown as dry spiny capsules and fall to the ground without, it seems, any means of dispersing the seeds. The durian, which is grown in villages and occurs also in the forests, has the largest fruits. They ripen greenish-yellow to light orange, and the large pale-brown seeds are thickly coated with white to pale-yellow pulp. This heavily armoured, tough fruit splits slightly when ripe and its five parts can then be forced open. The fruit smells of a mixture of onions, drains, and coal-gas, but the aril has no smell and tastes of caramel, cream, and, as some say, strawberries and raspberries. Usually the fruits detach when ripe and crash to the ground, where the pulp turns rancid in a day or two. In Malaya the smell of fruiting trees in the forest attracts elephants, which congregate for first choice; then come tiger, pigs, deer, tapir, rhinoceros, and jungle men. Gibbons, monkeys, bears, and squirrels may eat the fruit in the trees; the orang-outan may dominate the repast in Sumatra and Borneo; ants and beetles scour the remains on the ground. The discarded seeds begin to grow at once and establish massive seedlings under and around the parent trees. The great naturalist, Wallace, wrote in 1869; 'to eat Durians is a new sensation, worth a voyage to the East to experience' [83]. To contemplate the durian trees fruiting is a new sensation worth a botanist's while, who would endeavour to perceive what plant life has been doing.

Here is a modern tree competing in the high forest, yet provided with an archaic fruit without the means of distant seed dispersal. In this respect it resembles oaks, chestnuts, and many other large-seeded forest trees. Consider a vast land clothed with high mixed forest from the shore to an altitude of 12,000 ft. Of what use is distant dispersal, which, if it could be carried out, would merely remove the seedling to places where it might not be able to establish itself or take it away from others of its kind that could cross-pollinate it? It drops and lodges in the soil and with its own reserves establishes itself to regenerate forest when the older trees die. Lightly distributed seeds come into play when plants require wide dispersal to reach their sparser habitats. They belong not to the trees which regenerate forest *in situ* but to the pioneers of open places such as river-banks and landslips, to the epiphytes, and to the undergrowth, which must find bare places or quickly make use of sunny openings when a tree has fallen. The big seed is the forest tree's requirement and its powers of distribution are as limited as they are unnecessary; it is not subject to adverse conditions or to a prolonged dry passage that must be survived in a dormant state. Thus the seeds of most trees of tropical rain forest have no power of dormancy; they sprout at once. When dormancy develops by the drying of seeds in the ripening fruit, then such trees may extend their geographical range beyond the primitive domain. The small, dry, dormant seed, moved about ungerminated, has not the simplicity of primitiveness but the efficiency of experience. Contrast with the durian the seeds of willow and poplar, blown in early summer up and down the rivers in their longitudinal draughts to land on sand-bars newly exposed by the receding flood; or the grain of mustard tumbled out to roll over the soil, to become, maybe, incorporated and to germinate years later at the surface. The little fruit or the little seed, like the little flower, is the *multum in parvo* of long evolution.

To the complex of the arillate fruit the durian adds smell, rich taste, rich oily food, protection by spines while young, the immediate growth of tree seedling, and cauliflory or the faculty of flowering and fruiting from the mature wood of branch or trunk. These features occur in the arillate fruits of other families of trees but, like the aril itself, they have been variously degraded and lost in the evolution of other kinds of fruit. Traces of 'durian characters' are valuable guides, nevertheless, in understanding fruit biology, for they are bits of the inheritance and not, as often supposed, incidentals and ornaments produced by chance mutation. Thus the horsechestnut fruit (*Aesculus*) is clearly a small durian that has lost its aril but retained its spines, large seed, and fleshy rind: other members of its alliance (Sapindaceae) have arils and show variously, as in *Durio* and *Sterculia*, the substitution of the arillate fruit by dry, unattractive capsules, drupes, and winged one-seeded fruits (as in the maples). The great water-lily, *Victoria amazonica*, has a very spiny fruit, as indeed are all the underwater parts of its leaves, and the seeds have vestigial arils. There are palm fruits with knobs as the vestigial spines (*Teysmannia* and *Phytelephas*) or these are arranged in the armour of overlapping scales characteristic of the many lepidocaryoid (scaly nut) palms, to which the rattans (*Calamus*), the sago (*Metroxylon*), and *Raphia* belong. The knobs on some cocoa fruits (*Theobroma*) point to the durian spines in Sterculiaceae, even as the knobs on cucumbers are substantiated by the red spinous fruit of the dehiscent cucumber *Momordica* with modified arillate seeds. In fact, where arils occur spinous fruits may be expected, and where spinous fruits occur there will be other evidence of primitiveness in pachycauly: not even the spinous achenes of some buttercups can escape detection or, for that matter, the bristles of the gooseberry or the thorns of *Datura*. Thorns guard the new larder.

The pachycaulous state with massive stem and big leaves started the primitive broad-leafed tree that formed a massive flower, and this, according to the durian theory, developed into a massive arillate fruit, fitting the whole massive construction. Perhaps the undersides of the leaves were generally spinous. *Victoria amazonica* is such a pachycaul, debilitated by its aquatic escape from the forest struggle, but nevertheless primitively massive and spiny and surviving in the voracious backwaters of the Amazon. It is the character pre-eminent, too, in the scaly-fruited and spiny palms of swampy riversides. It is no coincidence that the only members of the orange family Rutaceae to have arillate seeds should belong to its spiniest genus *Zanthoxylum*. It is no surprise to learn that a cabbage tree of Hawaii, *Cyanea*, is spiny, or that spines distinguish the pachycaulous species of the tomato genus *Solanum*; and there are the red hips of thorny roses. Spines, in fact, are a part of the ancestral pachycauly of the flowering

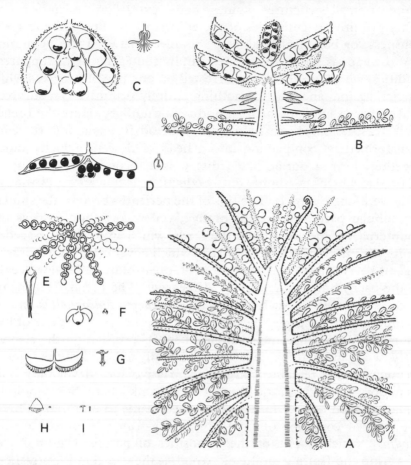

Figure 75. Scale of fruits and flowers; × ⅛. A, Ancestral pachycaulous spinous flowering plant, according to the durian theory, terminated by a structure both flower and fruit, the black seeds covered with red aril and ready for pollination and dispersal. B, Primitive fruit with spinous arillate pods, with detached stamens and petals and (in broken lines) the floral axis and carpels from which the fruit formed after pollination; note how the erect carpel or pod is unsuited to liberate the seed. C, Fruit and flower of *Durio*, inverted or hanging to let the seed drop out. D, Fruit and flower of *Sterculia*. E, Rows of plums developed from the many carpels of the flower of *Desmos* (Annonaceae). F, Hanging arillate fruit and flower of the nutmeg *Horsfieldia*. G, Hanging arillate pods and the erect flower of *Tabernaemontana* (Apocynaceae). H, Acorn and flower of the oak *Quercus*. I, Fruit and flower of Compositae.

plants and they point to the primitive state of the epidermis before it was restricted to the single layer of non-photosynthetic cells (figure 62). Then as trees, flowers, and fruits improved into the modern array of small structures, spines, hairs, extra-floral nectaries, and other primitive peculiarities of the epidermis disappeared and glabrousness, or botanical baldness, prevailed. Between the beginning and the end, as signposts on their family routes, occur the apparent oddities, which are the plants arrested on one line of progress while branching into another, for nothing entirely primitive can survive.

Another example of such an oddity is cauliflory. Here the bread-fruit trees *Artocarpus* can help again. A bread-fruit is an inflorescence transformed by compacting into a head of flowers, which, joined together, form a durian-like fruit: it is a second-order or super-durian, in which the minute one-seeded female flowers act as though they were single carpels. The tips of the perianths become the spines; the tubular parts become the orange, yellow, or white pulp as the counterpart of the aril around the seeds, which are really one-seeded fruits or nuts. This compound fruit, which never opens but works as a giant berry, must have come from an open-branched inflorescence in the pachycaulous ancestry of *Artocarpus*. The bread-fruit fits on to the stout twigs of such species as *Artocarpus anisophyllus* and *A. incisus*, but as the species become leptocaul, the fruit took one of two courses. Either it became small and kept its position on the slender leafy twigs, and in this case it seems generally to have lost the spinous character; or it remained large and massive and, developing only from dormant buds on the branches and trunk where there would be sufficient food reserve for its supply, it became cauliflorous. This is the familiar condition of the tropical jack-fruit (*Artocarpus heterophyllus*), which is one of the curiosities of nature. Cauliflory, or flowering and fruiting on the old wood, is the way that the advanced tree with slender twigs carries, like the durian, the ancient massive fruit that has not advanced. Then the fruit may advance and there come into existence trees such as the cauliflorous Annonaceae, *Baccaurea* (Euphorbiaceae), belimbing (*Averrhoa*, Oxalidaceae), or the ternate nut (*Macadamia*), which with berries or nuts give evidence of their route of evolution. In so doing, as Pijl has shown, they become suited to the ways of fruit-eating bats, which prefer to fly under branches or hang up on them while eating [84]. This is an example that shows how almost any deviation in the rich plant life of the tropics finds a counterpart in the activity of the rich animal life.

We do not know a pachycaulous durian tree, but experiment may make it, just as experiment may turn the pawpaw into a durian analogue.

One tropical family of trees, the nutmegs (Myristicaceae), has persisted with the arillate, though in their case one-seeded, fruit. They are advanced trees of willow habit with simple leaves set in two rows, but the twigs of some are stout, whereas those with slender twigs tend, as would be expected, to cauliflory. When pigeons gather soon after dawn to breakfast on the red aril or mace of nutmegs, the naturalist may witness the scene that terminates nutmeg progress. Looking into the past, as nutmeg trees become pre-nutmegs and begin to coincide, perhaps, with pachycaulous ancestors of the Annonaceae, so he will see pre-pigeons and primitive birds, perhaps *Archeopteryx*, gathering at the Mesozoic break of day. Then he may sharpen his wits on the durian and dream of pre-monkeys, pre-squirrels, pre-civet cats, pre-bears, and pre-bats gathering in clumsy confusion on low clumsy pachycaulous progenitors of durian, sterculia, bread-fruit, horsechestnut, palm, and banana, in the beginnings of flowering forest.

The story of the fruit, as that of the tree, leaf, seed, and flower, must be read in the tropical rain forest where life is unhampered by climatic extreme. The problem of the primitive seed in this vast and dense forest was not distant dispersal but the apparently small matter of how to become detached from the parent sporophyll. That animals did this originally is the gist of the durian theory. The two most primitive kinds of seed plants that survive are the cycads and maiden-hair tree, and their seed are distributed by animals. Previous to them seem to have been the fossil seed ferns, among which there are preserved numerous examples of seeds like those of the cycads, that is with fleshy, and presumably edible, outer layer and stony interior round the female prothallus. These seeds also seem to have been fully grown by the time that they were pollinated and fertilized. There is no escaping the conclusion that seeds brought the animals into the lives of plants. The impasse noted in *Selaginella*, when the embryo sporophyte had no means of dispersal, is overcome by the seed eaten by the animal. Then pollination was put back in the sequence of seed growth so that unopened seed leaves were pollinated on their tips, or stigmas, and pollen tubes grew down to immature seeds, as ovules. Into this process came in some uncertain way the primitive animals, which, in pursuit perhaps of the

223

primitive nectar of the pollen-drop or the exudation of primitive floral and extra-floral nectaries, inadvertently pollinated the primitive flowers. If the plants were spiny and unclimbable, then primitive flying creatures, as primitive birds or primitive beetles, may have reached the open reproductive buds, half flower, half fruit.

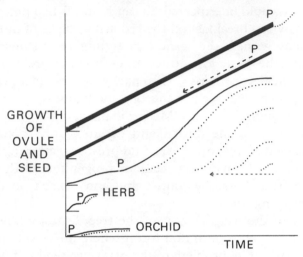

Figure 76. Derivation of the flower and the fruit shown by a graph relating development of the ovule, incidence of pollination, and development of the seed and of the embryo (dotted lines); P, point of pollination. Heavy line: as in cycads, *Ginkgo*, and presumably the primitive flowering plants, indicating the direct growth of the seed and its pollination when full-grown, the embryo developing after seed dispersal. Medium line: neotenic tendency of pollination leading to a flowering stage followed by a fruiting stage. Thin line: state in the large-seeded modern forest tree, with large flower and long post-pollination development of the fruit; the dotted lines showing the precocious tendency of the embryo, giving seeds with larger embryos. The herb with small flower, fruit, and seed. The orchid with ovule development subsequent to pollination, as the most neotenic state.

Then as the flowering and pollinating process was separated more precisely from the fruiting consequence of pollination, more orthodox creatures took over; insects developed pollination and vertebrates seed dispersal. There are still many botanists who seek to find in flower structure the primitive flower, which lies, of course, in the fruit from which by separation it evolved [85].

There is philosophic satisfaction in contemplating how the rooted plant, inventing the seed, made up for its immobility by employing the animal and brought about thereby the great and going concern that for some two hundred million years has been the era of flowering plants. This reproductive conquest changed the flora and brought in the tremendous development of the modern fauna that depends upon it; and civilization must be included so long as man needs plants. I am beginning to wonder if, without the note of vegetable beauty, the chord of humanity would ever have been struck.

Besides food from flower and fruit, the primitive flowering tree gave shelter in its branches. But the animals, inheriting as excessive reproduction, as usual overdid themselves, and flowering plants have since been fending them off. The new environment in the trees brought a flood of animals and the trees had to be protected. Green buds are camouflaged; scent and colour attract when the flower is ready. The green growing fruit is again camouflaged and prickled against inquisitive noses; it colours, smells, and opens when it is ready. Thus ripe redness became a factor in animal life, which ignored red seaweeds, and has its after-glow in our homes. Epidermal armour was broken through and reinforced with lignin. Woody fruits began to dry up, arils to disappear. Winged seeds, so often formed in the capsules which are most nearly related to arillate fruits, began wind-dispersal. Over-lignification shut the fruit. A pulpy cover kept it as an edible drupe, but thorough lignification produced the box of nuts to be seen in the Brazil nut (*Bertholletia*) and it has become a character of the family (Lecythidaceae), though it, too, has evidence of arillate ancestry. Progressive woodiness and loss of aril distinguish all the immediate allies of the durian, and with reduction in the number of seeds to the limiting case of one, true nuts such as the acorn and filbert have been evolved. Small seeds enabled the berry to be perfected as the unlignified indehiscent fruit that animals could gobble without crunching.

All this time, however, there must have been a chemical evolution, which concentrated in the unripe fruit such unpalatable substances as astringent tannins, bitter alkaloids, and sour acids, to disappear in the improved process of ripening. Chemical defence, of which we can do little more at present than sense the significance, took over from armour, and most fruits are now protected by chemical means from all the animals that would attack them, from boring weevils to rasping squirrels, destructive parrots, and inquisitive monkeys. Unless a young fruit be well camouflaged, armoured, or unpalatable, it cannot survive. Have you seen monkeys searching incessantly for food from sunrise to sunset and the splutter of objects that descend ?

Some trees have improved the primitive, rough-and-ready, dropping of their seeds by securing the embryo within the stone of the drupe or by means of the berry passing the small seeds unharmed through the animal body. Others have reacted from animal dispersal and have perfected the primitive method by dropping nuts, which may be washed or floated off by floods. Yet others have taken up, as with pollination, the wind dispersal of seeds or of winged nuts. In the meantime the great tropical forest has been built with its diversity of ways and living things. Branches, leaves, flowers, and fruits have been multiplied and arboreal animal life has kept pace, but the edible fruit supply, so intimately connected with the survival of the tree, has fallen off in safeguard. How all this has happened may still be read in detail from the inheritance and natural history of fruits, but biologists will have to hurry and take steps to preserve tracts of wild forest before ignorance obliterates them.

Chapter 13

The Fungus

TO THE FOREST-FLOOR, in the habitual excess of vegetable growth, fall twigs, leaves, bracts, sepals, petals, stamens, pollen, hairs, fruit husks, bud scales, bark scales, and discarded seed coats. Then the plants die: trunks and branches mingle with and compress the raw humus. Clean woods, as we have seen in Chapter 7, are an artifact. In the unexploited forests, now so distantly removed from civilization and becoming rarer so that most readers of this page will have not seen them and can have no idea of the quantity of debris, trees stand in all stages of dismemberment. There are gaunt giants killed by lightning, trunks supporting a few breaking limbs, unbarked shafts riddling into sawdust, which a nudge may topple, decaying stumps, and hollows in the ground where a few rotting roots still radiate. Other corpses are held in the coils of lianes; broken limbs dangle; one tree leans on to another; dead tufts of epiphytes hang precariously. There is as much dead matter standing and fallen as there is living, and it is not neglected in the economy of nature.

Mortality is inherited from the benthic stage of plant building, but it is not clear why or how a tree should die. It does not age uniformly like an animal but continues to thrust into the environment by means of growing tips and it maintains itself by secondary thickening round the moribund interior. Some trees can be propagated indefinitely, it would seem, by cuttings of twigs or branches. Intact, they appear to become too big to manage themselves. They grow to the limits to which water can be brought to the leaves and food to the roots. Five thousand years and 350 ft. above ground are roughly the greatest span of trees, but climbing palms with relatively slender stems 600 ft. long have been found. The redwoods (*Sequoia*) and the cycads hold the age record, redwoods and Australian *Eucalyptus* that of height, but for area covered the prize must go to the Indian banyan (*Ficus benghalensis*), for which, besides Alexander the Great's record, there is one of a tree in the Andhra valley with a crown that measured 2,000 ft. in circumference and was supported by three hundred and twenty root-trunks; it was as large as a modern factory, and well equipped.

In the tropical rain forest with the average tree canopy at 150–200 ft., giants have been recorded up to 260 ft., but their life span seems to be a mere three hundred to four hundred years. The crowns of the redwoods taper and peter out, like those of the papaya tree at 20–30 ft. when it is six to twelve years old. Palms and tree ferns peter out at heights that vary from 2 or 3 ft. to 150 ft. But the crowns of large flowering trees spread more widely and, when old, become stag-headed, as if the roots were dying, and this is certainly an important factor, difficult to unearth, in tree death. Small trees and shrubs, for some physiological reason, have generally shorter lives. Herbs can convert all their storage into flower and seed, and in one short season exhaust themselves. Yet this is not their peculiarity. There are palms, bamboos, and other monocotyledons like the century plants (*Agave*) that die after one gigantic fruiting; and there is a tropical American tree (*Sohnreya*) of the orange family which is said to grow to some 50 ft. high, to flower once and die. The age and the manner and cause of natural death in tropical plants are subjects very little investigated.

By contrast there are many reasons for trees being killed. Lightning strikes some. Others, top-heavy, fall over. Tropical storms blow big ones down. In falling, they knock over others or, roped together with climbers, collapse chaotically. So limbs are smashed. Wind snaps twigs and branches; heavy fruiting may do so; the weight of epiphytes may do so; the passage of animals may do so. The shading of lower limbs may kill them. For many reasons trees are wounded and in part die back. On to these dead surfaces come insects, which bore and chew to sawdust, and fungus spores, which grow into the dead tissue and decay it. Trees become infected and hollowed out until they are unsound – the pipe trees of foresters. They break up more readily and, though they may sprout with coppice shoots and replace the bole with a cluster of stems, once the rot has entered it will not stop. Fungi are the great rotters; they attack wounds with the minuteness of bacteria. New bark may form in time on clean cuts or abrasions, but splintered stumps cannot heal. Gums and resins can block such wounds with antibiotic dressing. Thus conifers and several kinds of broad-leafed tree, as the dipterocarps, may preserve themselves, but most trees are deficient. Even the copious latex of the giant fig trees is no avail and their light, quickly grown wood is most susceptible to decay.

Fungi are filamentous plants without chloroplasts. They are not therefore photosynthetic but live by decaying the tissues of green plants, less commonly of animals, and they can grow on the remains of any such plants except lumps of resin. The filaments are usually branching rows of cells $1-10\mu$ wide, covered with a firm wall made of chitinous or nitrogenous cellulose, but without cuticle. The partitions between the cells are walls or septa, which grow in from the side of the tube like a diaphragm closing until a small central pore is left as a communication between adjacent cells. Thus fungal filaments resemble those of seaweeds but their colours, when present, are never photosynthetic. The filaments are called hyphae. Their nature is insidious because, like fine pervasive fingers or hypodermic needles, they can insert in a microscopic way virulent chemicals. Through their walls they excrete enzymes that decompose the substance on which they are growing, and poisons that may kill, and through their walls they absorb water and the products of external digestion for their own growth. So, by internal pressure on their rigid walls, hyphae thrust their soft, growing tips forward to erode a passage into their source of food. Their externalized manner of growth into a surrounding food supply contrasts strongly with the contained growth of green land plants.

The weft of hyphae that forage form the mycelium of the fungus. The reproductive hyphae grow into the air from the mycelium that feeds them, twine round each other, and build the interwoven texture of the fructification on which the spores are produced. The white, yellow, red, brown, and black threads and webs on and in dead leaves, wood, and soil are the mycelium. The toadstools, puffballs, truffles, cups, brackets, clubs, and powdery spots are the fructifications developed by these mycelia. As land plants, fungi are dual and have immersed and aerial portions. Unlike photosynthetic land plants, the duality lies not between root and shoot as two vegetative parts, but between vegetative mycelium, complete in its saprophytic way, and the entirely dependent fructification, which disperses the results of vegetative activity in fungus spores. The fructification is a reproductive mechanism and becomes so toy-like in its ways, just as the fruit of flowering plants, that its study is also apt to be considered as a curiosity.

The variety of fungi is unbelievable. They have explored by means of their hyphae the dead world more effectively than any other kind of organism. In numbers they are the counterpart of insects. The explanation seems to lie in the limited ability of each kind of hypha. It is always a minute structure, which does not enlarge into a massive reactor with diversified tissues; thus it is restricted to the limited chemistry of its minute tube. Every species of fungus has its own chemical outfit and, as the species of green plants differ chemically and most of their component parts differ chemically, so there are as many species of fungus as there are species of flowering plant, if not of all seed plants, multiplied by the number of their parts. One species of fungus will grow on a dead rootlet, another on the wood of an old root, another on the old wood of a stem, another on a twig, or a piece of bark, the blade of a fallen leaf or its stalk, a fallen fruit, a

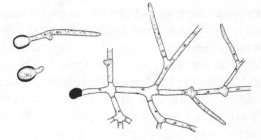

Figure 77. Spore of a fungus germinating into a hypha, which branches to build the mycelium in the food supply; × 500.

fallen inflorescence or a seed. Some with stronger mycelium course through the humus of damp leaves or through the soil with its detritus of humus, or in the mass of heartwood in a trunk. They take what their chemistry enables them to remove. They are followed by others which take other things, to be followed again and again, until the dead matter is rotted into detritus and a fine skein of effete hyphae, which insects, millipedes, worms, slugs, and snails devour. The centre of a hard trunk becomes a dark-brown soggy mass, which elephants, tapir, and rhinoceros discover; they snuffle up the natural cheese and blow the interior clean. The variety of fungi is a measure of the chemical differences in plant tissues. The size of the fructification is generally a measure of the extensiveness of the mycelium of the fungus, which in turn measures the extent of its food supply. The mycelium of the giant puff-ball travels widely among dead grass roots, spreading out on the periphery of a circle that may reach, in the course of two or three centuries, 100 yds. in diameter; that of a small toadstool may be limited to a leaf; that of a microscopic mould to a fraction thereof.

There are three sorts of fungus. There are entirely microscopic kinds that do not form fructifications compounded of many hyphae. Single hyphae grow into the air and end in a sporangium containing the small spores, 3–5μ wide. *Mucor*, the bread mould, with its black pin-head sporangium on a hypha, like a thread of silk, is the

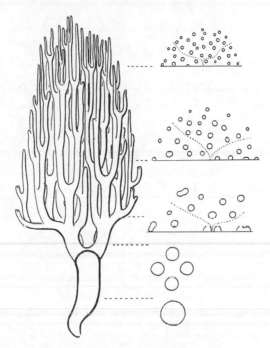

Figure 78. Fructification of the humus-dwelling basidiomycete *Clavaria*, with sections of its branches at various levels; × ½.

common example. The hyphae of these kinds, too, are rarely septate. They are called Phycomycetes, from their resemblance to the simple filamentous algae. In contrast there are those with fully septate hyphae, which aggregate in hundreds or thousands to form massive fructifications; and according to the manner in which these fructifications produce spores, so they fall naturally into two kinds. In one the spores are formed inside special elongated terminal cells, each of which is called an ascus, and it squirts out the spores when they are ripe. In the other group the spores are formed on very minute stalks on the outside of the end-cells of hyphae, each called a basidium, and the spores are flicked off the stalks one by one when they are ripe. Thus there are Ascomycetes (ascus fungi) and Basidiomycetes (basidium fungi), and to this second group belong all the toadstools, puff-balls, and large bracket fungi, from which it can be seen that they have also the most vigorous and extensive vegetative mycelia, exploiting humus, soil, and trunk [86].

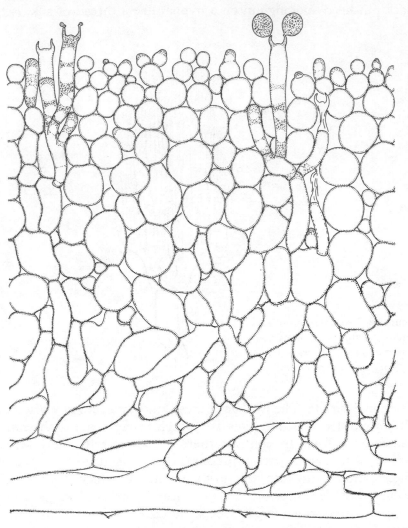

Figure 79. Outer tissue, or cortex, of the stem of a clavaria (*Clavulina cartilaginea*) to show its multifilamentous construction into false parenchyma as in the red seaweed; × 700.

Toadstools, puff-balls, and bracket fungi are by no means primitive forms of fructification. They are umbrellas and consolidations against environmental influence. By comparison with other fungi of the world's flora it is not difficult to see how they have come about. The most generalized form of the fructification in the basidiomycete is that known as *Clavaria* [87]. It is a massive branching structure arising from a stem that connects with the mycelium in the soil and humus. These fructifications, up to a foot high, variously white, yellow, brown, red, or purple, are quite common in undisturbed and primeval forest but are scarce where forestry is practised. They resemble the seaweeds compounded of many filaments, not only in general form but in microscopic structure. The cortex is not photosynthetic and the sporangia are converted into the basidia, which produce airborne spores, but they retain the externalized organization of the seaweed just as does the mycelium. This is the trouble of *Clavaria*.

The basidium is a sporangium perfected to work in air. It spaces the four spores that it produces most accurately round the apex. If it gets wet by rain falling on the fructification, its apparatus is destroyed; the top of the basidium collapses and the spores are not liberated. This is the weak point about *Clavaria* which has led to the toadstool. Instead of the branches growing erect, they are diverted towards the horizontal and, coalescing, they form the umbrella, which has a sterile wettable upper surface and a dry underside that is the fertile part. The underside is amplified into gills, tubes, or spines, which increase the spore-producing area and the spore output of the fructification. In bracket forms, which are elevated on the wood in which they grow, the stalk is done away with. The basidiomycete comes to tuck the reproductive part of its fructification underneath a cap. Then, with greater progress, it makes all this structure before it expands it by absorbing water. Thus the well-formed mushroom springs rapidly from a small button which takes many days to grow from the mycelium. If the cap fits too tightly on the button, it cannot expand; the gills or tubes within are thrown into irregular folds, and the puff-ball is born. Its spores cannot be flicked off to fall between gills or down tubes into the free air for dispersal, but the interior of the puff-ball dries up and through a new opening at the top the dry spores stream out as smoke. There are many kinds of puff-ball and their allies, the stinkhorns, and they are variously related to different kinds of toadstool, inasmuch as they

233

are over-elaborate toadstools that have gone wrong. By such alterations
in the direction of growth of branches of *Clavaria* there come about
the remarkable mechanisms of fungus fructifications, so sensitive to

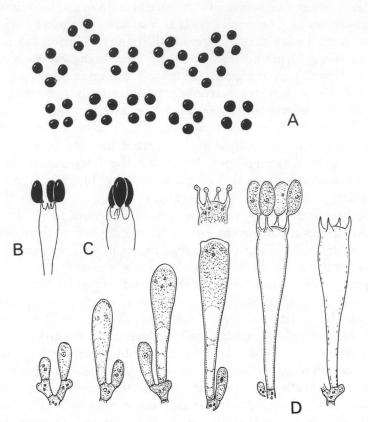

Figure 80. Spore spacing and basidium shape in the toadstool. A, Groups of
four spores on the gill of *Coprinus*; ×250. B, Basidium of this *Coprinus*,
showing the spores and their spikes aligned in parallel; ×250. C, Basidium
when wetted, collapsed and unable to discharge the spores; ×350. D,
Development of a basidium showing the nuclear fusion followed by meiosis into
four nuclei and the extrusion of the spores with the nuclei; ×500. (A, B, after
Buller, 1909–50, vol. III.)

gravity and light, which Buller has explained in his researches [88].

As for the life-cycle of these fungi, it is simple because it has been
greatly reduced. The basidiomycete has lost all its sexuality in reproductive organs but it retains, just as the flowering plants, the

sexuality of nuclei. The young basidium contains two nuclei. They conjugate and the diploid nucleus immediately undergoes meiosis to form four haploid nuclei that enter one into each of the four spores.

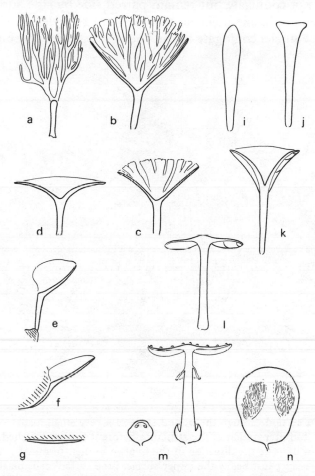

Figure 81. Two ways of making the umbrella basidiomycete: a–d, from the branched *Clavaria* form by webbing of the branches as they incline to the horizontal; i–l, by expanding and hollowing the top of the club of a simple *Clavaria* form and the development on the underside of gills decurrent on the stem (k), then free from the stem (l) as the elongating organ; e–g, the degeneration of the umbrella in wood-inhabiting basidiomycetes through steps with lateral stem (e), the sessile (f), and the resupinate (g); m, the perfected toadstool, as *Amanita*, with a button stage, in which stem, umbrella, and gills are made, and the expansion of the button to make the basal cup (volva), the ring (annulus), and the warts or flecks on the umbrella; n, the making of the puff-ball through loss of organization in the button stage of the toadstool.

235

When the spores germinate into hyphae in the soil or in some vegetable remains, branches may connect two hyphae from different spores, and a nucleus from one enters the cell of the other. The nuclei do not conjugate but remain paired side by side and together they divide and make mycelial hyphae with binucleate cells. Eventually these nuclei conjugate in the young basidium. Presumably the

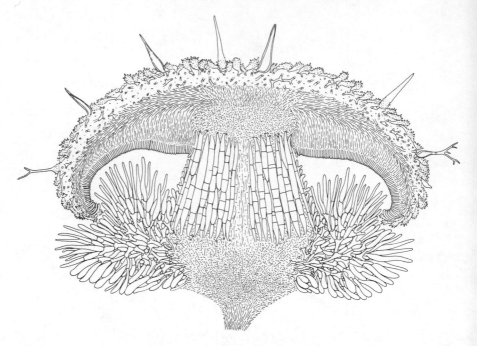

Figure 82. A section through the button stage of a very small toadstool *Mycena* to show the full preparation of the structure before it expands to shed the spores. Lines indicate the direction of the hyphae to their terminal cells; there is a mucilage layer just below the upper surface; the surface cells on the stem and the cap are called cystidia; × 250. Compare figure 83.

two nuclei that pair and later, after many simultaneous divisions, conjugate, are essentially male and female, and the basidium is the meiotic sporangium that produces the spores as zoospores adapted to air dispersal; but the prolonged binucleate state without conjugation, as almost the diploid state, yet not quite, is most peculiar and conceals a long history of very intricate fungus evolution, to consider which is outside the scope of this book.

236

Branching fructifications like that of *Clavaria* occur also in the ascomycetes, where they are as relatively infrequent. Their construction is similar. By multifilamentous growth they form a core of longitudinal filaments, or hyphae, covered by a cortex of oblique or divergent hyphae, the cells of which are short and swollen, as in the cortex of seaweeds, but never photosynthetic. They produce asci in

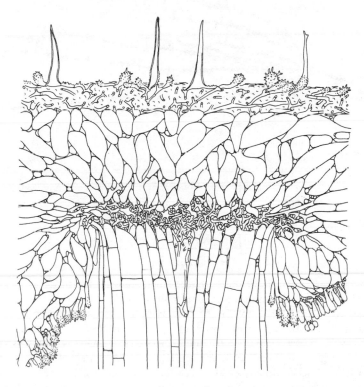

Figure 83. Central part of the *Mycena*, shown in figure 82, now fully expanded; × 250.

one of two ways, on which character the ascomycetes are divided into two kinds. In one, called Pyrenomycetes, the asci are produced in minute flask-like structures, each called a perithecium; they resemble the reproductive conceptacles of *Fucus* or the cystocarps of the red seaweeds. In the other group, called Discomycetes, the asci are borne side by side in open patches, which may be disc-shaped, saucer-shaped, cup-shaped, or ear-shaped through elongation on

237

one side. Such a collection of asci is called an apothecium, and the truffles are puff-balls of the discomycetes, which in over-elaboration have closed up. In both kinds of fructification there are copious mucilage-hyphae between the asci and in this mucilage the asci develop, as it were, under water. Asci therefore are truly aquatic

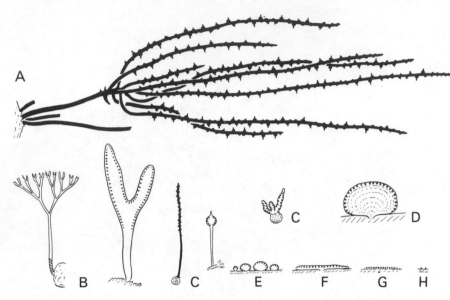

Figure 84. Fructifications of the pyrenomycete, showing the progressive loss of sterile tissue and the diminution in size; natural size. A, Seaweed-like *Thamnomyces* with black stems bearing many perithecia, growing on tree trunks in tropical forest. B, C, Various forms of *Xylaria*, on dead wood, fruits, twigs, and leaves. D, Black cushion of *Daldinia*, becoming an unplant-like body of minimum surface. E, Small cushions of *Hypoxylon*. F, Crust of *Ustulina*. G, Cluster of perithecia directly on the mycelium without organization into a plant body. H, Commonest form of pyrenomycete, as single perithecia on stems and leaves.

sporangia and their aerial adaptation consists in squirting out the airborne spores.

The ascus is also the cell where meiosis occurs. Like the basidium, it contains two nuclei that unite and immediately undergo meiosis to form four haploid nuclei, but then each nucleus divides again and around each an internal spore is formed. Thus the ascus produces eight spores and shows itself as less reduced than the basidium, which has arrived at the limiting number of four.

In yet another way the ascomycete shows its antiquity compared with the basidiomycete. When the hyphae that bear the asci are traced back to their origin they are found to come from a special fertile hypha (ascogonium), or one particular cell of it, and there may be one or several such structures for every perithecium or apothecium. This structure recalls the carpogonium of the red seaweed that produces the filaments which end in the tuft of carpospores. The

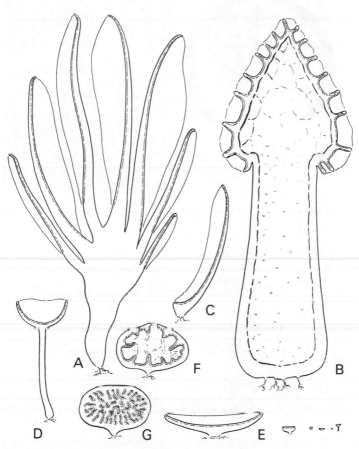

Figure 85. Story of the cup fungus (discomycete), shown by sections of fructifications, the fertile surface indicated by a broken line; natural size. A, Large humus-dwelling *Wynnea* with many ear-shaped lobes from a common stalk. B, Morel *Morchella*, supporting a honeycomb of cups on the end of the hollow stalk. C, Single ear-shaped cup of *Otidea*. D, Stalked cup of *Cyathipodia*. E, Sessile cup of *Peziza* and the common small and minute cups of its allies. F, Cup that defaults and by losing grip of itself closes up into the truffle (G).

239

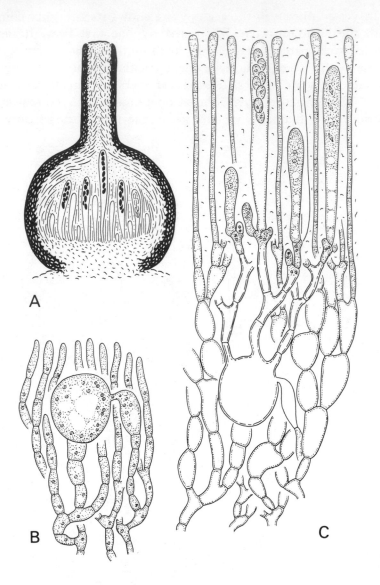

Figure 86. A, Perithecium in section to show the dark wall and delicate interior; × 40. B, Ascogonium of a small discomycete pierced by the antheridium, the nuclei from which are pairing with those of the ascogonium; sterile hyphae investing the sexual organs as the beginning of the fructification; × 500. C, Derivation of the asci of the discomycete from the ends of the ascogenous hyphae grown from the ascogonium, with nuclear fusion and meiosis in the young ascus to give eight nuclei around each of which a spore forms, the asci immersed in mucilage from the sterile hyphae (paraphyses); × 500.

cells of the special fertile hypha have one nucleus each, as have all the cells of the sterile hyphae of the mycelium and fructification of the ascomycete; thus they are unlike those of the basidiomycete with paired nuclei. The fertile hypha becomes binucleate by the entry of a nucleus into it from an adjoining cell, and this cell may be part of another hyphal branch, which resembles a male organ or antheridium. Then in its binucleate condition the fertile hypha forms short

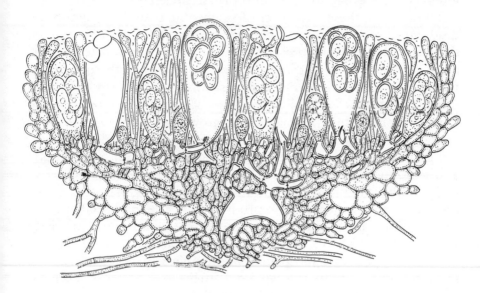

Figure 87. Section of a very small discomycete fructification *Ascophanus granuliformis*, showing the ascogonium and ascogenous hyphae in the central tissue; × 250.

hyphae with binucleate cells that branch and bear the asci at their tips. The ascomycete presents a life-cycle which resembles in essence that of a seaweed. The haploid gametophyte is represented by the mycelium and the fructification. Sexuality is extremely reduced but a fertile filament becomes binucleate in a manner of fertilization *in situ*, and from this is developed a parasitic diploid state resembling the carposporophyte of the red seaweed, though it differs in having paired nuclei and in producing meiotic sporangia with eight spores (octospores, not tetraspores, nor monospores as the carpospores are).

The problematic binucleate stage is brief and purely reproductive in the ascomycete. In the basidiomycete it is introduced early in the life-cycle so that both vegetative mycelium and fructification are made with binucleate hyphae. It may be that this sort of diploid state has led to the more vigorous mycelium and the larger fructifications of the basidiomycete. On all counts, however, the basidiomycete is the more advanced of these two principal kinds of fungus, but that does not mean that the basidiomycete has been derived from the ascomycete any more than that the fern has been derived from the moss. It is necessary to think into distant ancestry.

The point of these intricacies of fungus life is to show that in the dead world of land plants there are saprophytic organisms, unlike the green plants of the forests where they grow and unlike the animals, yet constructed and reproducing essentially in the manner of filamentous seaweeds. In detail the ascomycetes conform most with the red seaweeds, in which, because of fertilization *in situ*, the diploid sporophyte is a small plant growth parasitic on the free-living gametophyte. It is the case, too, with mosses and liverworts, but they, as parenchymatous plants, belong to a much higher grade of organization.

We are back at the teaching of the seaweeds. Their story tells how the organization, shape, and reproduction of the many-celled plant was the outcome of photosynthetic cells combining into one body attached to the shore, worked upon and affected by the moving water, and inheriting the planktonic methods of cell construction and reproduction by motile gametes. When similar features occur in land plants living under such utterly different conditions, they are evidence not of a parallel evolution on land but of the level that their sea ancestors had reached before they came on to land; the features are the sea heritage which the landward ancestors brought. The case of the fungi, living in the stillness of decay so remote from the commotion of the wave, is the clearest of all. How then could the seaweed have been changed into a fungus? This may seem an academic question, purely theoretical, but the answer is so simple, profound and revealing that it must be considered if there is a desire to understand, not just to know about, plant life.

Fungi now grow in situations created mostly by flowering plants and conifers. They are advanced habitats when the history of land vegetation is considered. As the mind travels into the vistas of the past, forests of seed ferns, tree ferns, and clubmosses succeed each other until there is a vague and messy picture of plants half in, half out of water becoming the origin of land vegetation. This morass created the environment of the saprophyte. It lay beyond the grasp of waves so that the corpses of plants stayed where they subsided. Under this mass, into it, and on top of it must have grown the root systems of the incipient ferns and mosses of the coming vegetation, but also there would have been all kinds of primitive seaweed growth which failed later on to survive in competition with the

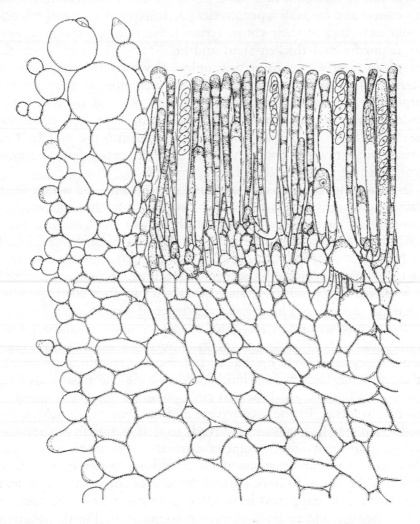

Figure 88. Section through the margin of the discomycete fructification *Coprobia granulata*, showing the enlarged cortical hyphae, the mucilage paraphyses, and the asci, resembling an opened cystocarp of the red seaweed; ×250.

243

better equipped. Among these, surely, were filamentous and multi-filamentous seaweeds, the very filamentous nature of which lent itself to insertive growth in the accumulating mulch. Saprophytism is known throughout the variety of existing land vegetation, from the colourless liverwort (*Cryptothallus*) and the gametophytes of clubmoss and adder's-tongue fern (*Ophioglossum*), to forest orchids. There can be no doubt that some primitive landward seaweeds took this course and became saprophytic, particularly as many, if not all, would have had in their construction those downgrowing hyphal-like filaments that thicken stem and hold-fast, which grow within the plant tissues by insinuating apical growth and which, being internal and non-photosynthetic, derive their nourishment by absorption from neighbouring cells. Such plants had hyphae ready made for saprophytic growth; they had only to escape from the parental plant body. Their spores fell on and into the mulch. The filamentous growths that they formed could easily have become hyphal growths and started the first mycelia.

Instead of joining the upward struggle for light, for which their filamentous construction was much less suitable than the stouter parenchymatous, a large section of the primitive landward flora would have become the undergrowth in the raw humus and rivalled the bacteria in saprophytic decay. They would have inherited according to their kind, which was the level of organization they had reached as seaweeds, a reproductive state wherewith to start the life-cycle, exactly as the seaweed inherited the essence of reproduction from its preceding plankton existence. Microscopic methods remained microscopic, as in the Phycomycetes, but the more massive multifilamentous seaweed body or thallus became the fructification or reproductive mechanism of primitive ascomycete and basidiomycete. The spores became air dispersed and sporangia became adapted to this air dispersal. In the ascomycete and basidiomycete, sexual reproduction may have advanced already to fertilization *in situ* and the asexual spores may have become non-motile, as in the red seaweeds. The ascus in particular shows itself as an underwater sporangium. Finally there seems to have ensued the loss of the organs of sexual reproduction, as happened in the flowering plants, but the essence of sexuality was also retained in nuclear conjugation. For the solution of the binucleate problem of the hyphae, it will be necessary to consider far more deeply the manner of hyphal growth in filamentous seaweeds.

With little effort of imagination it can be seen that on the origin of the land flora, there were opportunities for filamentous seaweeds to become fungi. Their new features would have been loss of photo-synthetic activity and, at the same time, the development of seaweed hyphae into a vegetative mycelium. Then the thallus, or compact seaweed body, which was inherited, became the reproductive mechanism fed entirely by the mycelium. If this was not the origin of fungi, then it must be explained first what happened to the filamentous seaweeds at the origin of the land flora; secondly, how nature dealt with the new environment of plant corpses; and, thirdly, why the fungus resembles so essentially the filamentous seaweed.

Chapter 14

Fungus Progress

BEFORE THE FUNGUS can form its fructification, its vegetative hyphae must forage widely. The food material is then sent back along their tubes to one or more places where reproductive hyphae are being made. A mushroom, for instance, does not send hyphae into the soil from the base of the stem, as if they were roots; it is the concentration from the threads in the soil that makes the mushroom. By what means this happens is far from clear, but it seems that food is pushed as cytoplasm from cell to cell along the hyphae through the pores in the septa as a result of the enlargement of watery vacuoles in other cells. Thus Buller has recorded cytoplasmic streaming through a length of more than a hundred cells of a hypha. When food is being concentrated, the reproductive hyphae are made in thousands. They are marshalled into columns, which are the stouter threads or cords of the mycelium. The columns grow to the surface of soil or wood and begin to build the fructification. All this time the columns of reproductive hyphae have to be fed from the mycelium and, even when the fructification expands, it still has to be supplied with water to enlarge the cells, though it probably carries enough reserves of food to make the spores.

A large fungus needs much material and much time to fulfil its nature. Its mycelium must grow where humus is rich and deep, or where there is plentiful vegetable matter, as in tree trunks. It must fruit when there is prolonged rainy weather and abundant water in the soil or tree trunk for the saturation of the hyphae in the fructification; wilted they cannot grow, and the daily drop in humidity as the sun rises may cause checks in growth, which appear as the zones on the upper side of bracket fungi. In fact, the chief stimulus to fruiting in modern fungi is heavy and prolonged rain after a period of moderate drought, which has checked the vegetative, or mycelial, growth. The soaking of the dry soil revives the hyphae, which proceed directly to form the fructifications. That is why in temperate countries there is the autumn crop of toadstools after the rains at the end of the dry summer. If a dry June in a northern country is followed by a wet July, then the early autumn crop may be advanced by a month or more. The late autumn crop of fungus, however, seems to depend also on lower temperatures and to be more or less unaffected by summer weather. In the ever-wet tropical rain forest, hardly a toadstool or puff-ball may be seen for many months, and this apparent scarcity has led to the belief that this forest lacks the abundant higher fungi of the temperate regions. Nevertheless, if one is able to stay

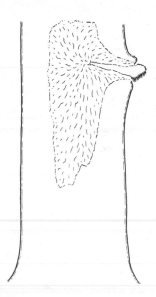

Figure 89. Living beech trunk (*Fagus*) occupied by the mycelium of a coral fungus *Hericium erinaceum* decaying the heartwood and producing its large fructification through the rotted centre of the stump of the branch by which its spores probably gained entry; $\times \frac{1}{24}$.

long enough to observe the forest frequently throughout the year, the same regime may be detected. There will come a short, sharp, dry spell when the humus crackles underfoot. Then, even in an hour after the downpour, the new fungus growth can be smelt, and after a week or ten days a crop of quickly grown fungi will appear and continue for a week or two, as if all the fungi of the northern August and September had been reviewed in a fortnight [89]. I have been in the forest in Pahang when it has been impossible to tread without stepping on earthstars (*Geaster*, figure 91), eleven or twelve kinds of which had suddenly appeared as if from nowhere. On these occasions the monkeys come down from the trees, for they know well the tastiness of mushrooms, and snatch from the pigs to the satisfaction

247

of the tiger. The tastiest fungus in the Old World tropics is the toadstool *Termitomyces*, the mycelium of which is cultivated by the white ants, or termites, in their underground nests. This fungus does not occur in the New World tropics, where, in strange contrast, the leaf-cutting ants (which are true ants) cultivate another and quite unrelated toadstool, referred to a little-known genus (*Rozites*), but possibly a true mushroom *Agaricus* [90].

Figure 90. Fructification of the clavaria *Scytinopogon* growing up from its mycelium deeply situated in the soil (shown in black); natural size.

It follows that large fungi cannot fruit and therefore cannot survive on small bits of dead matter such as a stick or two, a leaf, a fruit-husk, or even an animal dropping; that they cannot survive in poor soil with little humus; and that they cannot survive where drought prevails. For such places the fungus must have a small mycelium and a small fructification to fit it. Small size enables the fungus not only to use small opportunities but to fruit more often on small occasions; it has a shorter life-history. The fructifications of a

248

large *Clavaria* may take a month to form and another month to complete sporing. A small toadstool or indeed a small *Clavaria* just a few millimetres high can complete its whole life-history from spore to spore in a much shorter time on a single leaf; some, in fact, need only a week. The spore output of the small fructification is, of course, much less than that of the large, for the surface area of the fructification is a ready measure of its spore output. But, having so many more little places in which to grow, it can produce many more fructifications, which, in the aggregate, maintain the spore output needed to overcome wastage.

The extremes of fungus life are the slow massive growths with large spore output and the small quick growths with small spore output, which give the quick returns where massive growth is impossible. These are the ways in which ascomycete and basidiomycete display themselves. There are relatively few genera of massive construction or relatively few species with massive construction in the various genera, and there are quantities of genera and species with small or minute construction. In all those cases where direct comparison can be made, as for instance in such large genera as *Clavaria*, *Polyporus*, or the mushroom itself (*Agaricus*), the peculiarities of the small species are seen to be those of the large ones simplified to fit the reduced habitats. It becomes clear that fungus evolution has proceeded from extensive massive growths to smaller and simpler growths that fit the multitude of brief habitats such as the fragments of vegetation afford: in other words, the simpler fungi are derived from the more elaborate massive ones. Body construction is no longer the competitive factor in their lives as it must be in the upward struggle of photosynthetic plants. It can be dispensed with very profitably. So the common fructifications of cup-fungi (discomycetes) take the form of little cups, saucers, and goblets, a few millimetres wide, with the minimum of tissue to support the asci. Those of pyrenomycetes reduce to single perithecia, less than a millimetre wide, and all the sterile stalk tissue is not needed. In very poor circumstances they can survive by producing one very small fructification; in good circumstances they can produce cheaply many small fructifications. The basidiomycete is more complicated because of its more effective mycelium. Thus it has improved the big *Clavaria* into the big toadstool, bracket fungus, and puff-ball, but among all these kinds there is the same tendency to excel in little forms. The smallest and most reduced in both main groups of fungi are the microscopic

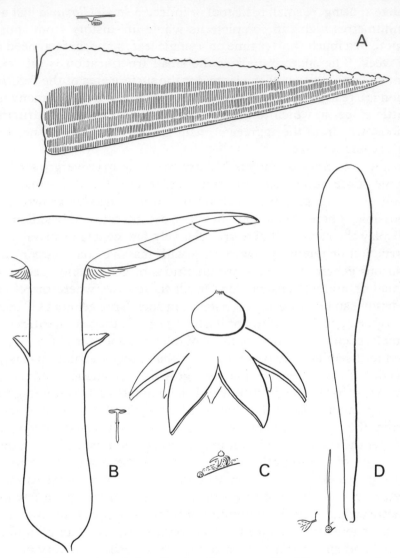

Figure 91. Large and small fructifications in genera of basidiomycetes; × ½. A, Large perennial pore fungus *Fomes senex* of the tropics, with five seasonal layers, growing on a dead trunk, and a minute ally growing on the midrib of a living leaf. B, Large mushroom *Agaricus*, forming wide fairy rings in the soil, and a small ally growing on small patches of bare soil. C, Large humus-dwelling clavaria (*Clavariadelphus*), its ally *C. juncea* on a twig, and the related *Typhula* on the stalk of a dead leaf. D, Large humus-dwelling earth star *Geaster*, and its small ally on a stick.

250

yeasts; they live in such brief places as rotting flowers, rotting fruits, and slimy exudations from wounds, and they have become little more than spores budding and perpetuating spores. The green land plant has built upon its structural heritage, but the fungus has progressively eliminated as much of it as possible [91]. A great part of the fungus world comprises therefore microscopic forms that appear as flecks and stains on the dead and living parts of plants.

It might be supposed that the remains of mosses, ferns, club-mosses, and horsetails provided the first humus for fungi and that therefore they should still harbour more primitive fungi than seed plants. This is not the case. The comparatively few fungi that grow on such cryptogamic remains are merely small and simplified allies of those that grow on conifers and flowering plants: nor do the cycads and *Ginkgo* help. The fungi of today are those of coniferous and flowering forest that adapt themselves to other kinds of vegetation. Ferns and their like do not decay rapidly. The thickness of the Coal Measures, composed of the remains of fern-like plants, indicates that even at this stage of the world's history there were not many organisms able to remove their detritus. Sphagnum bogs are another example. Conifers, too, with resinous tissue are not so easily decayed as flowering plants. It seems indeed that it could not have been until their advent that the fungus world expanded. Flowering plants employed the animal and propagated the fungus. The flowering forest is both eminently edible and putrescible, wherein lies the other half of its success. Its remains do not accumulate. They are rotted down by fungus, animal, and bacterium and reincorporated in the soil. The flowering forest has developed a cycle in which material for plant growth, mainly water and mineral salts, is absorbed by the roots, built up into the plant tissues, and then discarded in dead pieces on to the forest floor, where they are rendered again into the state in which they can be absorbed by the roots in the top-soil. No coal or peat can start here. By continuous turn-over, the flowering forest maintains itself. Intact it prospers. Cut down, cleared, and burnt off, the soil is exposed, beaten by rain, washed away, then dried out, beaten and washed off again until gullies appear on its bare surface. The exposed soil is overheated, over-dried, and over-illuminated for the soil organisms accustomed to shade; its life dies. After the top-soil has been exploited, it cannot be regenerated. Where stood the grandest forest, there may be in a few years of extravagant cropping truly waste land.

Top-soil is the gold in the basement of the forest. It is the micro-cosm where, to repeat, the lower forms of animal thrive with fungus, bacteria, and roots; they work on the fallen remains of plants and animals as soon as they become available. On the plant side, in addition to cellulose and lignin, which are reconverted by soil organisms into sugars, there is the remainder of dead protoplasm in the plant-cells, and the use of it is one of the more recondite aspects of living chemistry [92].

First in tropical soils come the termites. In armies from earthen barracks they speedily organize on their six-legged carriages batteries of jaws which fracture and comminute the material; in their insides they can digest lignin. They generally work at night, when the munching can be heard. Thus trunks may vanish before fungus or beetle has been able to work on them. Then, various insect grubs, woodlice, millipedes, snails, and slugs rasp at the remains that the termites have left; earthworms tug pieces into their burrows where they soften by bacterial action and are worked eventually through the worm. The fungi grow more slowly on all and sundry, but their chemistry is far more varied and they reduce to edibility such material as bark, which is not at first palatable. The animals excrete, defaecate, and leave slimy tracks. The fungus excretes and leaves tracks of threads. Bacteria multiply on these. Protozoa multiply on the bacteria. Rotifers, nematodes, and other minute animals multiply in these soil colonies and add their excretions. Beetles, spiders, centipedes, and scorpions prey on the soil animals and add their quota of waste products, all of different chemistry. Fungus spores abound, germinating and ungerminated, and on these microscopic eggs many protozoa, rotifers, and mites feed. A mouldy smell, compounded mainly of the many minute fungus odours, marks the solid, liquid, and gaseous decomposition in this microcosm of chemical engineering.

Raw humus varies much in depth. The collapse of a tree with its tangle of climbers and epiphytes may create a heap of decomposing matter. A litter of palm leaves may be several feet thick: a litter of *Eucalyptus* bark, which is slow to rot, may accumulate for 8 or 10 ft. Where, however, raw humus consists mainly of the fallen leaves of trees it is merely a few inches deep. Loose at the surface from fresh ingredients, it is beaten down by rain and the passage of animals, and it collapses as it disintegrates until it becomes almost laminated in fine layers of progressive decomposition, the ultimate particles of which in minutest fragmentation mingle with the upper layer of the sub-soil.

The main roots of trees grow into the sub-soil for anchorage and support. Branch roots end in feeding rootlets, which grow into the top-soil where they act as drains to remove the products of decomposition for resynthesis in the tree. Some kinds of tree are assisted, it seems, in this matter by an intimate connexion with the hyphae of certain soil fungi, and the two growing together form what is called mycorrhiza. The hyphae cover the rootlets with a fine

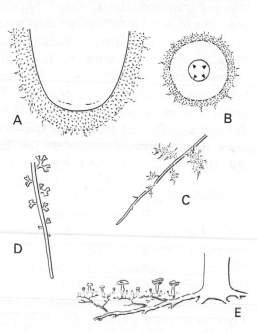

Figure 92. Mycorrhiza. A, Apex of a beech root in section, showing the mantle of hyphae; × 160. B, Section across a small beech root mantled with hyphae; × 20. C, Rootlet of a broad-leafed tree with mycorrhizal side rootlets in dense clusters; × 2. D, Rootlet of a pine tree with dichotomous mycorrhizal rootlets; × 2. E, Base of a tree with mycorrhizal toadstools springing from the roots; much reduced.

mantle of threads or, in other cases, the hyphae enter the rootlet and without detriment establish themselves in the cells of the cortex. This close connexion of fungus and rootlet seems to be essential both for the fungus and the tree. Oaks, beeches, birches, and their allies, and many conifers have mycorrhiza and the fungi associated with them are some of the various kinds of toadstool and truffle characteristic of their forests; the fungus mycelium mantles the rootlets and spreads in the humus. In other cases, as the heaths and rhododendrons, the fungus is microscopic and seems to live only in the cortex of the rootlet. To what extent mycorrhiza occurs in tropical forest is not known but it certainly accompanies the great variety of oaks in the Asiatic and Central American tropics, and from the constant association of certain toadstools and truffle-like fungi with others it may be a fairly general occurrence. The fact is that the inextricable root world of the mixed tropical forest is too complicated and difficult for any ordinary study.

How exactly the mycorrhizal association works is by no means clear, and the most recent investigations show that there is no simple relation between the components [93]. Nevertheless it reveals the manifold methods by which trees in mixed forest recover nutrient material from the mixed humus. No one kind of tree, any more than one kind of soil organism, can deal with all the humus and it seems that there are as many methods as there are species of root. Each extracts what it can with the help, direct or indirect, of animals, bacteria, and, above all, fungi. Conversely one kind of tree with its particular chemistry is only one way of exploiting soil, air, and sunshine. The more the kinds of tree, the greater the exploitation, and the richer is the dependent animal and fungus life. Above ground, green plants compete uniformly for light and carbon dioxide; thus they battle and the forest heightens. Below ground they have many more requirements. They compete, certainly, for water but soil chemistry offers many more opportunities for diversification than air chemistry. The roots of some species will compete for certain mineral and organic materials, but those of others will pursue their specialities; most, perhaps, will be involved or will co-operate in working trains of chemical events from which they will severally take their needs. Thus roots intermingle and the root layer of the forest nowhere extends as deeply into the earth as the canopy into the air. There is no comparable downward struggle. Nevertheless, above and below ground flowering forests have developed the fullest world of life. What they have done and, to some extent, how they have done it are clear. The origin of the flowering plants, however, being a mystery, so is the origin of the animals and the fungi that they attract, because it is difficult to see how such could have existed in any of the fern-like forests, which are known, geologically, to have preceded the flowering.

Fungus spores are extremely small, 1–10μ wide, rarely as much as 15μ. They are produced in enormous quantities and shed mostly into the air. A giant puff-ball will produce seven million million spores; a large bracket fungus – the largest may reach 8 ft. in diameter – may produce tens of thousands of millions: a mushroom will produce a few thousand million. Spores therefore are ubiquitous [94]. Most are carried up into the general circulation of the atmosphere and die: their very smallness so lessens their rate of falling that the least draught sustains them. Strong light and prolonged desiccation kill them. Even if they survive long passages they will probably come down where they are unable to grow. The wastage is the most enormous in the living world, and we have yet to assess the contribution of these fungus 'eggs' in the food chains of animal life. The spores of small fungi growing in sheltered humus will be more likely to fall on suitable ground but, if too effective, many spores will oppose each other like a cluster of seedlings in one spot. There is a happy medium, to secure which the great majority of spores dispersed in the most random manner must be lost. We may without hesitation assume that the survival rate for spores is far less than one per total output of the spores from a fructification.

Among the lost spores are those that rain continually and unperceived on to living plants and animals. The dry, tough, chemically inert cuticle of the green parts of plants is arid enough to stop most spores from germinating. Likewise the bark, mechanically and chemically, is a barrier. In animals, some property of the skin such as the hard and inert chitinous cuticle of invertebrates, the slime of frogs and toads, the dry scales of reptiles, or feathers and hairs, stop and filter off the spores. Internally animals may digest them, occlude them with mucus, or chemically knock them out with antibiotics. Fungi, with their limited chemistry and easy access through the absorbing hyphal wall, are particularly susceptible to chemical warfare; they wage it on each other. Rarely for instance do fairy rings, which result from the outward growth of mycelia, cross each other. Usually a leaf, a stick, a branch, or a large part of a trunk produces one kind of fungus fructification, indicative of the exclusive growth of one species of fungus. It is first come, occupy all, because the spore on germinating, and then all its subsequent hyphae, exude in traces chemicals that can stop the growth of other species.

Nature, however, is artful. Many small fungi, derived through other habits, overcome the living defences. Their spores can send fine tubes through the cuticle of green plants and, on entering the living cells, proceed to kill them with antibiotics, and to thrive and spore on the dead and dying tissue: or the hypha from a spore enters through a stoma. This is the parasitic habit of mildews, rusts, smuts, and other small fungi that grow on living leaves, green stems,

255

flowers, and fruits. Generally their powers are limited and they attack only certain species of host, or, indeed, varieties of a species, as happens with the cereal rusts. Hyphae that attempt to enter unsuitable hosts are killed by antibiotics when they come in contact with the protoplasm of the host. Thus it is that parasites rarely get the upper hand in nature. Their spread is impeded by the array of unsuitable hosts that surrounds them. With cultivated crops spread is easy; spore wastage is reduced because the spores may blow immediately on to an adjacent host; and fungus diseases become the pests of agriculture and forestry. Most perfect among such parasites are some of the smuts, so called from the sooty powder of their spores. They may infect seeds, then seedlings, and grow inside the host behind its growing point until the host, which is outwardly normal, produces instead of pollen or ovules the mass of fungus spores; the parasite is the master chemist.

In other cases fungi are chance parasites. Of those that usually grow saprophytically some can attack living plants and kill their tissues by entering through wounds. This is the manner in which many fungi attack the trunks and roots of trees, and become epidemic in orchards, rubber estates, and forest plantations [95]. It is the habit generally of basidiomycetes, the extensive mycelium of which can cause a mass attack. It is the habit, too, that some think may have led to the origin of mycorrhiza, in which the host tree has become the master of the fungus.

Nor do animals escape. Various microscopic fungi attack hair, horn, feathers, scales, skin, and even chitinous cuticle, causing skin diseases. Most insidious to their small hosts are the so-called entomogenous fungi of insects. Their spores enter by mouth or air-tubes and grow inside the body, which is gradually and numbingly converted into a mass of hyphae inside the mummified skin. Through a weak joint, the finger-like fructification of the fungus emerges. Many such mummies can be found in the forest where the spore content of the air and the incidence of disease are high. There are also microscopic filamentous fungi in the soil which capture rotifers and small nematode worms on sticky hyphae, or trap them in hyphal coils; then they send a hypha into the animal body to digest its contents; such fungi are both predators and parasites. Higher animals seem not to suffer so much in the wild but, domesticated, they may encounter the mass action of spores, as when mouldy fodder is eaten. Then moulds, which are normally harmless, overcome in their quantity the antibiotic resistance of the host, and lungs, intestines, kidneys, and womb may be infected. In human beings, fungi are generally limited to skin diseases, more prevalent in the humid tropics as microscopic and almost yeast-like growths where the dead parts of the skin and hair join the living.

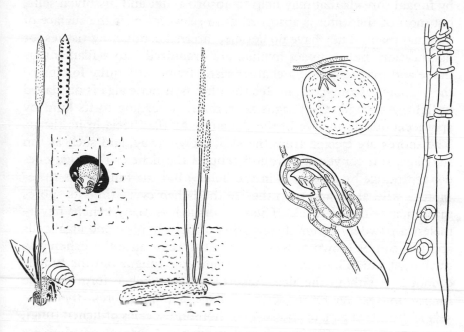

Figure 93. Fungi on animals. Left, the pyrenomycete *Cordyceps* growing on a beetle grub in rotten wood, and on carpenter bees, one of which has died at the mouth of its hole in a dead branch; natural size. Right, predaceous soil fungi, one of which (*Stylopage*) has captured an amoeba and the others have captured in their coils nematodes; *Trichothecium* with small nematode, *Dactylella* with large; ×250 (after C. Drechsler, 1937; C. L. Duddington, 1951; A. J. Juniper, 1954).

By happy contrast, the fungus has been tamed by the alga. There are many kinds of single-celled and filamentous green and blue–green algae that have migrated to land and live in rain water on soil, rocks, and trees. Thus, like mosses, they have acquired the faculty of drying and reviving again on wetting. Some of these become wedded to fungi, chiefly to the ascomycetes. The fungus spore begins to grow in the same sort of place and, if its hyphae encounter a suitable alga,

17

they surround it and incorporate it, without inconvenience, into their mycelium. Among the hyphae the algal cells grow and multiply, and they influence the hyphae so that they form not a diffuse mycelium but compact, tough structures, which in their irregular lobes and leaves, branchings, and stems, resemble small seaweeds. They are the lichens of the land. They have no roots, though they are attached by fungal threads that may help to absorb water and dissolved salts, but most of the water is absorbed as in seaweeds over the surface of the dual plant. They have no cuticle, internal conducting tissues, or lignification, but compact hyphae are organized into a filamentous tissue and made to build a characteristic framework quite foreign to the ordinary ascomycete; in this the photosynthetic alga is displayed to light. At length the fungus reproduces according to its kind by apothecia on the surface of the lichen or by perithecia in its tissue. The spores are ejected from the asci, blown away, and such as can start again in convenient ground rebuild the lichen. The alga does not reproduce by zoospores in the lichen, but the fungus has to recapture wild algae and resynthesize the lichen every time its spores recommence the life-cycle. There is formed by many lichens nevertheless a powdery material consisting of both hyphae and algal cells and this can be blown or washed about to reproduce the lichen as a dual organism. The alga can live without the fungus but the fungus cannot live without the alga. According to the lichen form that the fungus makes, its reproductive organs, and its spores, there are several hundred genera and several thousand species of lichen fungi, but comparatively few genera and species of lichen algae: they are shared by the fungi [96].

The lichen is the best example of the dual organism of symbiosis in biology. Its two components form a plant which has the form neither of the fungus nor of the alga but of a resynthesized seaweed modified to the land with algal cells immersed in a filamentous cortex. By virtue of the thick walls of both fungus and alga, the double plant can withstand drying. Lichens thrive, however, with great luxuriance in wet places where they are not overgrown by other vegetation, such as the rocks of the mountain cloud-belt or the rainy sea coast, where in forms of seaweed and sponge they give the aspect of a raised beach, or in smaller growths on the poorest soils of wet land before they have been colonized by larger plants. Nevertheless they can grow in arid places with an intermittent water supply, as on desert rocks and the trunks, branches, and leaves of trees. In this

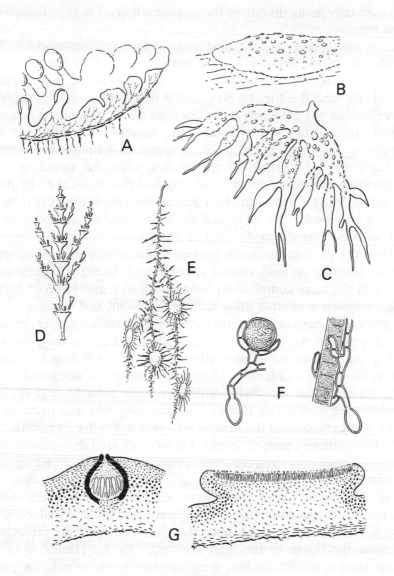

Figure 94. Lichens, natural size. A, *Peltigera*, attached to the ground by an elaborate system of fungal roots, and bearing apothecia at the ends of its lobes. B, *Lecanora* encrusting bark. C, *Ramalina* hanging from a branch. D, *Cladonia* with tiers of cups rising from the ground. E, *Usnea* hanging from a branch. F, Fungus spores forming thick-walled hyphae and encountering algae; × 500. G, Sections of the perithecium and apothecium of lichens to show how the fungal hyphae (short lines) build the structure and retain the algal cells (black) in the sterile parts.

259

capacity they make up one of the characteristic, if minor, features of land vegetation.

How the partnership is conducted is far from understood. The algal cells exude presumably the carbohydrate excesses of photosynthesis that the fungus must use, because it has no extensive mycelium, and the fungus may exude nitrogenous and phosphatic excesses that the alga may use. The alga will oxygenate the fungus tissue during photosynthesis and the fungus may supply carbon dioxide to the alga. On the whole it seems as if the fungus benefits by the food that it obtains from the alga, while the alga directs the fungus to house it in a much better manner that it could do by itself. Recent investigations with the electron-microscope reveal that the fungus can probe the algal cell with extremely fine processes of fungus protoplasm into the wall of the alga.

How the partnership arose is even more mysterious. Theories of chance meeting on land, chance success, and chance co-operation to result in the large complicated lichen with so many seaweed factors beg the question of what those factors represent, and how they could be evolved under land conditions. Such theories, moreover, mistake the degenerate and even the disintegrating lichen for the primitive. The point is that the photosynthetic cells of the lichen are single-celled or simple filamentous algae which have not advanced to the structural level of building multifilamentous growths and forming hyphae; yet they act as if they had, for they take the place of the photosynthetic cells of the fungus ancestor and order its hyphae into multifilamentous seaweed shapes. Church referred the association to a recombination of the primitive fungus of the period of migration of plants to land with the minor algae of the transmigration [97]. Thus the vision of land plants may be threefold. The green parenchymatous plants built the land vegetation. The larger filamentous plants decomposed it. And those that failed in both ways associated to make the lichen as the land seaweed. We are reminded of the three states of planktonic life, mentioned on p. 30, namely the plant, the animal, and the plant–animal. On migration to land, the plant continued, the fungus arose, and the lichen was compounded.

Chapter 15

Beyond the Forest

HIGH RAINFALL, sunshine, and temperature make the tropical forest the prime of plant life. Plants have never been able to work dry methods. By origin aquatic, they have inherited this need for water and, even on land, grow best where its supply is unlimited. But stagnant water is not enough. Roots prefer daily rain oxygenated by the atmosphere as the tidal substitute. Then, through photosynthesis and making of cellulose, the more the sunshine the bigger is the frame that the plant builds. Warmth increases the rate of all plant activities, but warmth with high rainfall is always tempered from excess: even on cloudless days the shade temperature in the rain-forest climate seldom reaches 38°C. So, under the optimum set of conditions, the flowering plant has entrained fungus and animal to build in those fortunate parts of the world the finest forest; and to the whole extent of this primeval tropical forest the traveller Humboldt gave the name hylea, taken from the Greek word for aboriginal woodland [98].

As soon as plants go beyond the hylea, circumstances for their growth become difficult. Conditions become exacting; the plants need to be specialized; the vegetation that they compose begins to become characteristic. Rainfall lessens and the end is desert. Trees become deciduous; twigs are thin and dry, or thick and fleshy; desert plants tend to be leafless, spinous, or succulent. Seasons enter and lead to winter. Days lengthen, nights shorten; then days shorten and nights lengthen. Periods of prolonged activity are followed by periods of inactivity, imposed upon the plant in its vegetative state and while it is a seed. Temperate lands have the spring flush of leaves, the early summer flowering, the autumn fruiting, and, between times, the miscellaneous flowering, which is determined by the succession of lengthening or shortening days. Prairie, misty mountain-top, and tundra have their character as water, light, and temperature approach the minimum for plant growth.

The one thing that cannot be said about the hylean forest is that it has character. Its plants have, of course, specifically or generically, but as a whole it is far too manifold to be singular. Trees differ side by side, as do the epiphytes and climbers on them. One may be leafing, another flowering, and another fruiting. On the ground are palms, bananas, burrowing gingers, ferns, and dicotyledonous herbs of perennial or short-lived habit as different as they represent so many families. The ground flora has no uniformity whatsoever; nor have the trees. There are big trees and little trees. There are big-leafed trees with small flowers, and small-leafed trees with big flowers, as there are leafy parasites with small flowers (the mistletoe family), and leafless parasites with enormous flowers, as *Rafflesia*. Some leaf continually, others intermittently; yet others are deciduous. There are all kinds of flowers and fruits. There are big seeds and small, which sprout at once, and others that are dormant for various periods from a matter of days to a sleep of years. In the hylea the flowering plants have produced almost every variation that seems possible on every part of the original theme, and these variations fit into the growing complexity as a forest. It is difficult to appreciate because so little has yet got into the books about the lives of hylean plants. But these variations were the set-up for extra-hylean conditions [99].

The epiphyte on a branch, accustomed to water shortage, is fitted to grow on rocks and, by further accommodation in water storage, to grow in deserts. The great family of bromeliads, as *Bilbergia* and the pineapples, have this adaptability from hylea to rainless desert in tropical America; the lichen-like *Tillandsia* may, indeed, clothe telephone-wires, where it can live on dust and dew. With less amplitude orchids, Melastomaceae, aroids, and ferns commute between the epiphytic and the rock habitats. In contrast, there is the tree form of tropical oaks with intermittent growth of the bud. It opens and develops flower spikes. When it opens again it develops leaves, which photosynthesize while the acorns are growing. When they ripen, the bud develops more flowers, then leaves, and so on. If winter arrests growth, the evergreen oak flowers in spring and fruits in autumn, or the deciduous oak bears flowers before the new leaves are fully-grown in spring, wind-pollinated in the absence of insects, and sheds its leaves when the acorns have fallen off. The temperate oaks employ the equipment that tropical oaks have developed in the hylea. A tropical *Hibiscus* tree will begin to flower after eight to fifteen months from seed and continue daily. If it can grow in a seasonal climate, sprouting in spring with lengthening days it may flower earlier and fruit in one summer as a growth of one season; if its root-stock persists, it becomes a perennial herb such as a mallow. Tropical hollies (*Ilex*) seem merely to be slowed up in their evergreen state by the temperate winter; in fact, deciduousness occurs in some tropical hollies of lowland and swampy hylea. Just because many plants of

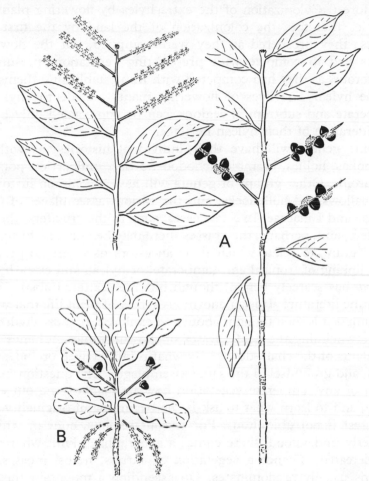

Figure 95. Leafing and flowering in oaks (*Quercus*); ×¼. A, Tropical oak (*Q. conocarpa*, South-East Asia) producing by intermittent growth a leafy shoot followed by a flowering shoot with scale leaves; then, with the next set of leaves, the acorns ripening and the old set of leaves falling off. B, Common European oak (*Q. robur*) with this habit adapted to the leafless winter, spring flowering with male inflorescences soon falling off, summer leafing, and autumn fruiting.

temperate climates are deciduous, it is presumptuous and teleological to suppose that the habit was developed there for that purpose. The history is that plants in a new environment had to make do with what they inherited.

What has come to survive therefore in places outside the hylea is what was pre-equipped before it left. Plants that inherited suitable faculties improved and prospered outside the stiff struggle in the rain forest. Colonization of the extra-hylea by flowering plants resembles therefore the colonization of the land by the first land plants; they prospered as they were equipped, but the flowering plants had to compete with pre-existing vegetation, presumably coniferous, as they had competed with it in establishing themselves in the hylea. The lives of flowering plants outside the hylea, in temperate and subtropical regions, can be appreciated only by the consideration of their hylean heritage.

Some genera will have their individual histories, as with the magnolias, hollies, brambles, rhododendrons, mulberries, potatoes, and aroids. Other groups of genera will have a common history, as the mallows and hollyhocks, and the various assemblages of Compositae and Liliaceae. In a few cases, such as the crucifers, the umbellifers, and, perhaps, the grasses there may be only one history for each family, to relate which their ancestors have left no progeny. This linking of tropical and temperate, or hylean and extra-hylean, botany has scarcely begun, though bits are scattered abstrusely in scientific literature. It is an enormous aspect of plant life that will be lost unless it is soon thought about and explored, because civilization is overwhelming the wildernesses of the subtropics. Whence came the plants of the chalk-down? By what routes arrived orchid, vetch, grass, and groundsel in the same assemblage? The question may be asked of any temperate vegetation how its plants came out of the hylea, and to learn what to ask is the beginning of originality.

Forest is not ubiquitous. For various reasons, none of which is properly understood, there comes a geographical limit where trees are defeated. Thence a vegetation of shrubs, herbs, mosses, and lichens usually predominates. On ascending a mountain the trees become smaller until they are stunted and sterile and, finally, disappear; shrubs then prosper with herbs, and only mosses and lichens may persist at high altitudes. The causes are numerous and intricate; the heights where these changes occur vary with latitude, aspect, and, indeed, the size, isolation, and rock of the mountain.

Temperature falls on ascending and both the early morning minimum and the early afternoon maximum are important for the first
may kill and the second may be insufficient for photosynthesis:
plants differ much in the temperatures at which they will form
chlorophyll and at which it is active. The rain is colder and the soil
and soil water are colder. Roots may differ from shoots in their
temperature needs. It may be hot enough in the day for the leaves to
photosynthesize yet the soil may be too cold for the absorption of
sufficient water; or the roots may absorb while the leaves cannot
work. Soil temperatures in the hylea will vary from 24° to 30°C,
and soil temperature, rather than air temperature, may be a better
measure of forest change with altitude until frost sets in. On a
tropical mountain lowland tropical forest gives place at about
4000 ft. to forest with a temperate, but evergreen character, having
smaller, leathery leaves and, often, needle leaves; thus, laurels,
myrtles, oaks, heaths, rhododendrons, and conifers predominate as
in subtropical and warm-temperate regions at 30–40° latitude. The
same species may occur on tropical mountains as in the subtropical
zones but generally there are specific differences in accordance with
differences in climate.

Then sunlight increases in intensity up the mountain, and strong
sunlight has a dwarfing effect by preventing the lengthening of internodes. Winds increase and will alter conditions of evaporation and
shoot temperature. Cold drying winds will shrivel young shoots,
stunt growth on the windward side, and generally cause bushy and
flat tops by accentuating the growth of side shoots. Probably the
most important factor at high altitude is the thinning and im-
poverishment of the soil. It is washed away by rain, crumbled away
by frost and drought, and brought down by falling rocks and leaning
vegetation. Towards the summit there is less hold for the roots and
less for them to grow in; furthermore water drains more rapidly or
is stagnant in pools. There is neither enough soil nor, in dry weather,
enough water for large plants to survive. So trees are defeated, unless
they can suffer dwarfing to prostrate bushes a few inches high, which
happens often on tropical mountains. Under conditions of extreme
altitude other trees, particularly the conifers, may establish seedlings
that root in fissures of the rock, but they remain permanently
dwarfed and sterile: such are the miniature conifers of Japanese tray-
gardens which may live as dwarfs for two or three centuries. The
same effect of summit-denudation can be seen on limestone hills,

merely 500–2000 ft. high in the tropics; the lower two-thirds are clad with luxurious forest and trees may root down for 200–300 ft., especially the fig trees; but upwards the vegetation gives place to orchids, grasses, and drought-resisting ferns.

On the Himalayas, broad-leafed tropical forest extends to 3000 ft. Above this the evergreen subtropical forest extends to 6000 ft. Higher, the coniferous forest of pine, spruce, fir, larch, cedar, juniper, and yew extend to 12000 ft. There follow ericaceous scrub, alpine meadow, and, lastly, the moss and lichen on bare rock with a few dwarf flowering plants in sheltered fissures. Then there is the latitudinal parallel, first emphasized by Humboldt in his Andean exploration; outside the tropical hylea, there is subtropical forest at 30–40° latitude, coniferous forest at 40–60° latitude, to be followed by the ericaceous scrub of moors and finally, the subarctic tundra of grasses, mosses, lichens, and relatively few other dwarf flowering plants [100]. Mountains at different latitudes are scaled accordingly, and in the Northern Hemisphere, northern slopes show lower zonation than the warmer southern slopes.

The main factor in latitude zonation is winter with its lower temperature and light-intensity, but the most decisive factor is frost. Slight frost may scarcely affect subtropical vegetation, but frost of considerable intensity or duration kills off those plants that have no means of overcoming its effects. It is not easy to draw a frost-line on the map because in many subtropical parts of the world, for instance Florida, frosts severe enough to kill back the branches of tropical trees or the leaves of tropical palms may occur only at intervals of several years. Tropical plants grown in such places could not withstand such die-back in competition with subtropical, hardier vegetation: in fact, the natural vegetation itself is always the best measure of world climate. Rubber trees of the genus *Hevea* delimit the American hylea, dipterocarp trees that of Asia. Palms, Bignoniaceae, and the caesalpinioid and mimosoid trees of the Leguminosae may indicate the frost-lines. Maples and horsechestnuts, as sapindaceous derivatives, and oaks and beeches carry the vegetation zones to the coniferous belt, where the pines must imply some climatic effect. Further north, for there is no corresponding land in the south, shallow-rooted conifers such as the Siberian larch (*Larix sibirica*) survive where the sub-soil is permanently frozen until they peter out into dreary Arctic tundra suited to the long night of winter, the continuous day of summer, and above all intense frost.

On muddy tropical coasts, and their muddy estuaries or lagoons, there develops directly the mangrove forest with its gloomy character [101]. About the limits of the tropics, mangrove trees peter out. The colder air and water of winter seem to stop them. They cannot withstand frost and they have no subtropical or temperate counterpart. In extratropical waters therefore their place is taken by the characteristic saltmarsh herbs as the sea lavenders, sea asters, sea chenopods (beets, *Salicornia*, *Obione*, *Sueda*), and the seagrasses [102]. On the sandy beaches, however, even in the tropics, there is often a belt of herbaceous plants as grasses, sedges, beans, and convolvulus or

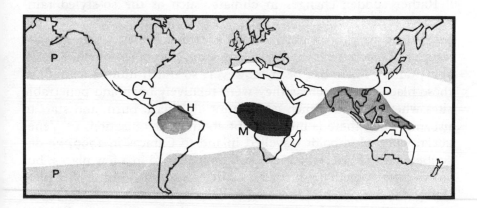

Figure 96. Distribution of palms, P, indicating the botanical frost line; of para-rubber trees (*Hevea*), H, indicating the hylean Amazon; of dipterocarp trees, D, indicating the hylean Indo-Malaysia, and their allies (Monotoideae), M, indicating the hylean Africa.

morning glory in front of the forest. Tree seedlings sprout in this herbaceous belt and, in a few years, will establish young forest, which will shade out the herbs. If the coast is advancing, the herbaceous belt will be maintained and followed by the forest along the seaward front. In this case the herbaceous plants display a technique that is employed in most places where new ground is being planted; they do not replace the trees or necessarily pioneer their entry, but get in before them.

Where forest verges on a dry climate, the trees become shorter, more often deciduous in the dry season, and spaced further apart. Those that maintain themselves develop strong root systems, which

267

exploit the water supply of the soil effectively for considerable distances and prevent thereby the establishment of a close stand. In the intervals, shrubs and herbaceous plants fill the niches both of soil and season when rain falls. With drier climate trees may fail entirely or their place is taken by succulent trees as the cacti and century plants in America and the spurge trees (*Euphorbia*) and aloes of Africa; spiny shrubs also enter and others with deeply rooting, woody underground parts which may tap the sub-soil water at great depths; then, when rain falls, seeds sprout, herbs spring up and blossom in their rapid way, and roots may push deeper to the underground water.

Rather sudden changes in climate, such as the so-styled rain-shadow areas of mountains, may transform forest as rapidly into desert landscape. Slow changes that accompany the passage into the interior of a continent involve a long transition of savannah, steppe, and prairie. There has been, however, so much human activity in these places, inasmuch as they were relatively open and penetrable sites where primitive men could hunt and herd, burn, and start to urbanize, that there is much doubt about their causation. In Venezuela, when Humboldt travelled inland to Caracas in 1799, he described dense hylea, which can still be imagined in a few places, but wood-cutters, charcoal-burners, cattle, goats, and wanton fire have since denuded much of this area to barren rock on which the cactus and spiny mimosoid trees of the arid coast scarcely survive. On Singapore island, in the hub of rain forest, there are hills which have been so ruinously cultivated for a hundred years that they are derelict 'lateritic deserts' without top-soil: lumps of resin from dipterocarpous trees can still be found in the earth as proof of the luxuriant forest that formerly grew there. By such means human activity can turn the richest vegetation under the same climate into desert; it is false therefore to suppose that climate controls vegetation, for hylean vegetation ameliorates climate. In the British Isles, with salt-laden air, soft rain, rare thunderstorms, and an antipathy to goats, we can form no idea of the destructive power of heavy rain or the protective power of dense forest, no idea of over-grazing where trees, for lack of rain, are disappearing; and no idea of the effect of burning. The Mediterranean region, the Middle East, and most of India and China are the effect of civilization over a few thousand years. Some go as far as to maintain that all desert, even the Sahara, is the consequence of early and destructive civilization, which removed trees and admitted over-grazing: even alpine meadows are artificially extended grazing [103]. Yet the very existence of wild herds of cattle, sheep, goats, antelopes, and horses, summed up conveniently if not precisely as ungulates, in Central Asia, Asia Minor, Africa, and North America, indicates that there must have been primitive grassland. One or two thousand years of stone-axing and burning may well have extended the areas and developed the vast herds of the past centuries into a phenomenon of recent history, which with firearms, steel, and internal combustion we cannot expect to maintain. Nevertheless it does seem that, where trees were defeated by increasing aridity, falling temperature, natural fire, and, in consequence of all these things, poorer soil, the evolution of grassland has been the final and most dramatic contribution of the flowering plant to animal existence.

The comparison of a grass and a tree, of a meadow and a forest, is a sterile exercise until it is remembered that a tree, as a seedling, is also a small plant. Then the herbaceous plant is perceived as the incipient state of a larger and woody plant induced to flower precociously. It fits thereby, like the small fungus in its world, the brief or inadequate circumstances unsuitable for tree growth, gives quick small returns, plentiful enough in the aggregate where there is room for many plants. The barley harvest on poor, dry summer soil is the result.

The grass makes a good example to understand the herb. The leaves are narrow and thin; the stem is hollow and slender; the delicate inflorescence is set with spikelets on fine stalks; the flowers are minute, flattened, and simplified with few stamens on the slenderest stalks; they are wind-pollinated and contain a single ovule; the dry fruit gives the thinnest covering to the seed, replete with endosperm; the roots are equally slender; the plant is refined [104]. As a monocotyledon it has no secondary thickening. The stem size depends on that to which the growing point enlarges from the seedling state. In many monocotyledons, as mentioned in Chapter 9, the seedling stem gradually widens as it forms more leaves and becomes the stouter adult stem, built upon the same axis. In the grass, as in the orchid and sedge, the seedling stem bears a few leaves and then a branch from the base bears a larger stem with more leaves, to be followed by larger and larger stems until the final and specific size is reached. There is not one main enlarging axis but the grass branches into a tuft of enlarging stems, the biggest of which usually become

269

the flowering stems; thus it reverses the tree habit in which branching leads to smaller twigs. A few slender grasses can, however, flower on the seedling stem, for instance small species of *Phleum* and its allies; they make use of the briefest situations, but this is also the habit of the maize (*Zea*), which is in many respects a very unusual grass, the nature of which has not been realized in spite of much research.

As it is a monocotyledon, leaves of grass grow basally. They form

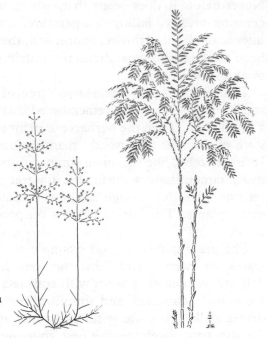

Figure 97. Grass with basal branching, and its terminal inflorescences with verticils of branches (without bracts); × ½. Bamboo with basal branching, scale leaves on its stout main stem, and distal branching with grass leaves; also, a grass plant to scale; × $\frac{1}{100}$.

sheaths that support the hollow stem and protect both the apical bud and the basal growing regions of the internodes situated just above the node. The construction is ideal for grazing because the biting of leaves does not injure the growing point or, if it should, then new branches can be formed from below and the sward thickened: the internodes can elevate the trampled grass, which invites and survives grazing. Grass that is ungrazed chokes itself with long dead leaves.

The grass seed has no special means of dispersal. Its small size enables it to be washed, tumbled, or blown about, or to be picked up with mud on the feet of animals, as cats and dogs convey the seeds of grass weeds about in the garden. The dry remains of the flower are shed with the seed (really a one-seeded fruit) and, when they have hairs or spiky points, they may stick to animal fur and be carried greater distances. The seeds can remain dormant for various periods and if, as often happens, they fall in places unsuitable for germination, they can await another conveyance. They are far different from the arillate forest seed, which germinates at once. Its size, immediacy, and security, have been exchanged for smallness, large numbers, delay, and uncertainty. Even the starchiness of the endosperm fits the character because oiliness, so common in arillate seeds, soon causes rancidity and the death of the embryo.

At one end of the grass family (Gramineae) there are the small spring annuals such as diminutive species of *Aira*, a few inches high, or the slender matgrass *Nardus* of acid and ultra-hylean moors. At the other end are the giant bamboos of tropical and subtropical lands which gave the timber for houses, the material for household goods, and the medium for writing in the civilization of south China. They are rarely encountered in the original tropical forest, unless there has been a landslip for their entry, though there are slender forest climbers among them. They belong rather to the outskirts of the rain forest where the climate is becoming seasonal and the forest less dense. They are, indeed, gigantic grass weeds of forest openings and, of course, clearings. The bamboo, as described in Chapter 9, has the same general construction as the grass but it is more massive and, as one would expect, in several ways more primitive so that the origin of the slender grasses must be sought in the ancestors of the pachycaul bamboo. Thus the bamboo flower may have a hundred stamens. Its inflorescence has bracts, and the stems branch freely. They produce several, even up to thirty, branches from one node in exactly the same manner as the grass tuft branches. Thus a branch is formed from the axillary bud; then from the base of the branch another forms to branch in its turn, and so on until there sits on the node a tuft of branches. The bamboo inflorescence branches in the same way, and the slender grass repeats this tufted habit in its economical inflorescence without the attendant bracts, which are unnecessary in a structure formed in the protection of leaf sheaths. The grass minimizes, as it is not a forest competitor.

Some bamboos of tropical Asia have large red plum-like fruits, several inches long, but without a stone, and this is clearly the antecedent in fruit evolution to the dry 'nutlet' of the grass. It is indeed the indehiscent modification of the arillate fruit of the bamboo ancestor, and such stoneless plums are paralleled in the arillate derivatives of other families of plant (for instance, the genus *Melia* of the mahogany family Meliaceae). The seeds of such fruits have no

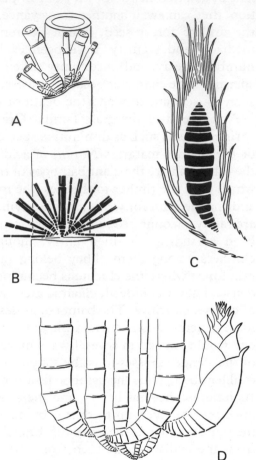

Figure 98. Bamboo branching. A, Node of the giant bamboo *Dendrocalamus* with a tuft of branches, the leaves and scale leaves removed to show the nodes and buds; $\times \frac{1}{2}$. B, diagram in explanation: the first branch (stoutest) produces side branches from its base; they branch from their bases, producing smaller branches, and so on. C, Basal bud (bamboo shoot) that will grow into a leafy stem; set with scale leaves and containing internodes (in black) already hollow and beginning to extend; $\times \frac{1}{10}$. D, Basal branching of a bamboo tuft, the new shoot on the right comparable with that of C, the roots omitted; $\times \frac{1}{10}$. (A, after Takenouchi, 1931.)

dormancy and germinate directly, even while hanging on the bamboo branches.

The bamboo seedling grows like the grass seedling and builds up in the course of years a tuft of larger and larger, increasingly pachycaulous stems, until the massive buds, 1–8 inches wide, of size characteristic of the fully grown state of the species, are produced.

The meadow grass corresponds with the bamboo seedling and represents a bamboo precociously matured in its slender state with slender inflorescence and small seed; it is a neotenic bamboo. Between their extremes come many grasses of intermediate stature, to be classified as grasses on botanical details but appearing more as bamboos. There are the fen reed *Phragmites* 2–12 ft. high, the Mediterranean reed *Arundo* up to 25 ft. high, and the Amazon reed *Gynerium* up to 60 ft. high. In other grass alliances come the tall sugar-cane (*Saccharum*) and the elephant grass (*Pennisetum*) with solid stems, more primitive than the hollow bamboo. The small meadow grass is, in fact, the end product of reduction and simplification by neoteny of many lines of grass evolution; among these derivatives must be put the dwarf bamboos, less than a foot high, in Japan.

Some bamboos, especially those with the large plum-like fruits, have the faculty called hapaxanthic flowering; that is, the individual plant flowers once and for all. Huge clumps of leafy stems are built up over periods of twenty to forty years without a trace of flower. Then they pass abruptly into the flowering state. The food material in the clump, from root to leaf, is entirely mobilized for flowers and fruits. The flowering starts on the leafy twigs and extends down the stems to the very base. It continues for several months, even for two to three years, and with the last fruits the clump dies. With this clump, too, have fruited and died all other clumps of the same species in the region, and any clump that may have been grown from a transplant of the same generation. The process cannot be stopped; death of the bamboo is inevitable, for any new shoot started at the onset of flowering that might serve for propagation, even when cut off and planted elsewhere, enters the flowering state. With so many seeds lying around, rats and mice multiply and when the bamboo seed is exhausted they attack the farmers' fields. The villages lose the bamboos, which serve so many rustic uses, and gain a plague of rodents, which may carry plague themselves. This has been going on periodically in eastern Asia ever since the bamboos originated, and it can be traced in historic annals by one manifestation or another, such as dearth, death, or failure to pay taxes, over a period of two thousand years [105].

Now this ability to mobilize all available material into one exhaustive fruiting is exactly the faculty of the grass that makes it, with slender construction, such an efficient seed producer. It is the faculty that the breeder selects for the maximum output of grain

18

with minimum straw. It is the faculty of the short-lived herb, which fruits and dies in its short season. We begin to understand how plants that are not herbs have the faculties which may enable their evolutionary descendants to become herbs.

The case of the grass is clear because there are so many examples in its large family to point the way; because such a big and close alliance bespeaks relatively recent evolution, perhaps within the last sixty million years of the Tertiary period. But the grass exemplifies the general principles of herb-making, whether monocotyledon or dicotyledon, and they can be listed briefly: (*i*) precocious flowering gives the slender state, corresponding to the seedling or juvenile state of the pachycaulous forest ancestor; (*ii*) inflorescences bear many small flowers giving many small seeds, often as small one-seeded fruits; (*iii*) seeds become dried and dormant; (*iv*) the life-history can be passed successfully in situations too brief for massive or extensive growth (sufficient small seed can result from such short vegetation).

There are many parallels such as the rushes and sedges, which can in no way be considered modifications of bamboos. There are the orchids, the gingers, the aroids, and all the herbaceous alliance of the lilies. Though generally similar in dwarf habit, they differ so much in leaf, flower, fruit, and seed, if not also in actual manner of growth, that they are clearly parallel herbaceous products of the several lines of simplification from a hylean ancestry. The banana is an example of the first step in herbaceous habit, still retaining pachycaul construction and hylean situation; it represents a state such as could be derived from the Brazilian *Phenakospermum*, but banana ancestry has mostly disappeared. Even pachycaulous palms and screw pines have their slender dwarfs suggestive of the precociously fertile seedling state. It seems that any kind of robust forest monocotyledon may have produced herbaceous forms if it had the faculty of precocious flowering, and this is to be interpreted perhaps in terms of the production of growth substances in the plant and the susceptibility of its tissues to their concentration. As small plants, the herbs fit into the regime of nature whether primarily as forest undergrowth or secondarily as plants that anticipate afforestation or proceed beyond it.

Among dicotyledons there are numerous and outstanding tropical genera that consist of large trees, small trees, shrubs, herbs, and, in some cases, even climbers, showing how a large part of the spectrum of plant form on land has been evolved within the limits of a single genus from the pachycaulous forest-maker. *Hibiscus*, *Cassia*, *Solanum*, *Vernonia* (Compositae), *Clerodendron* (Verbenaceae), and the complex of *Gardenia* and *Randia* (Rubiaceae) are examples. Some of the herbaceous forms relate, like the banana, directly to the pachy-caulous, others to the leptocaulous. It seems in fact that at any stage in tree evolution or of the derivative evolution of climbers and lianes there has been the tendency to throw off, as it were, herbaceous forms. Thus there are twining herbs in the Convolvulaceae. The strawberry is a herbaceous derivative of the raspberries and black-berries (*Rubus*) of the Rosaceae. The strawberry runner is but the herbaceous counterpart of the stout blackberry spray which bends over to root at the end; indeed the stout new stems of blackberries, covered with prickles, and containing a wide pith, appear as one of the primitive pachycaul states of the family.

Large numbers of papilionaceous herbs are related variously to woody forms, whether trees or climbers; for instance, the French bean (*Phaseolus vulgaris*) is the herb related to the runner bean (*Phaseolus coccineus*) in their climbing alliance, and the many tropical herbs of *Desmodium*, simulating clovers, are related to trees in their alliance. Saxifrages are herbaceous derivatives of the tropical tree forms of their family, milkwort (*Polygala*) of its family, violet of Violaceae, St. John's wort of Hypericaceae, and the purple loose-strife (*Lythrum*) of Lythraceae. None of these can be understood as herbs in temperate countries without considering their hylean derivation. We might as well try to relate modern art of modern countries without considering their historic character, as try to relate the herbaceous forms of temperate lands directly with each other.

In other cases there are pairs of allied families that stand to each other as the woody hylean ancestors and the herbaceous, which are mainly the temperate derivatives. Their affinity is so close that, if it were not for the custom of classifying them into two families, which has arisen from the growth of botany in temperate countries, the pairs would be united in blocks comparable with such uniformities as Rosaceae, Leguminosae, and Gramineae. The hylean Cappari-daceae relate with the herbaceous Cruciferae, Bignoniaceae with Scrophulariaceae, Verbenaceae with Labiatae, and Moraceae (the mulberry family, so incredibly varied in the tropics) with the Urti-caceae, familiar as nettles. The umbelliferous herbs (parsley, celery, carrot, hemlock, and so on) are particularly interesting because they relate to the pachycaulous hylean ivy family Araliaceae. The giant hemlock (*Heracleum*) has a short pachycaul stem and rosette of enormous leaves, as a 'stemless' pachycaul plant, and then like the hapaxanthic bamboo after several years it forms a giant inflorescence, fruits, and dies. Lesser hemlocks flower in the third or second year, and other allies of them are annual herbs, flowering from the small rosette of smaller leaves, as leptocaul herbs, in the first season; some, however, as carrot and parsnip, retain the pachycaul tendency in the edible root. In the mountains of East Africa and Socotra there are umbelliferous shrubby treelets (*Heteromorpha*) allied with the more familiar *Bupleurum*, and in New Caledonia, which is another island retreat, there is a genus (*Myodocarpus*) that combines the characters of Umbelliferae and Araliaceae. These are the plants, so little known or appreciated, at the edge of the forest or beyond the trees from which we may improve our ignorance: they point to the pachycaul hylean ancestor.

The dicotyledon repeats the teaching of the monocotyledon that, when herbs are seen in great profusion and variety in temperate lands, particularly in agricultural and urban community, their origin is not to be found short-sightedly in local circumstances but must be viewed distantly in geological time as the outcome of herb-making in and around the hylea. The herbaceous habit is the outcome of the flowering tree by which flowers and fruits have been carried almost to the ends of the earth. To understand needs a knowledge both of the herbs and of their hylean progenitors, such as we do not have, and this chapter cannot yet be written with any hope of satisfaction. The story of the groundsel (*Senecio*), which we do not know in any precision although the evidence must still be with us, is not the same as the story of clover, speedwell, buttercup, periwinkle, grass, or early purple orchid. Here are innumerable matters for research, fascinating and tantalizing, for they will reflect light forwards and backwards on the history of plants [106]. The curious botanist can never rest.

When a new kind of plant organization succeeds in the struggle for life it comes to dominate the earth; it specializes into the countless variety of situations; it displays, as zoology says, adaptive radiation, and it ousts its predecessors. The flowering plant diversified into tree families, which overwhelmed the previous forests and travelled on with herbs. They met the relics of previous herbs and so, among the extra-hylean meadows, there are horsetails, clubmosses, and adder's-tongue fern (*Ophioglossum*) to be found along with grasses, dwarf willows, and small heathers. How do they survive, surrounded with modernity? Perhaps their different chemistry enables them to root with grasses, for both will have different requirements and soil-extractions. In deserts, however, these archegoniate

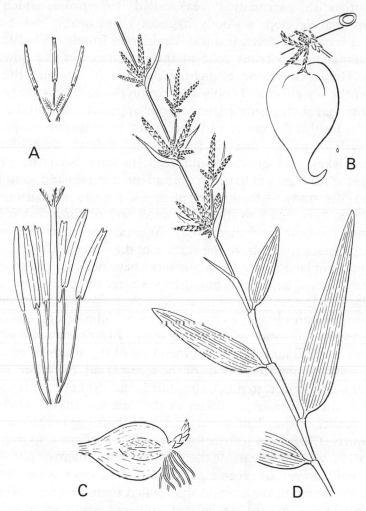

Figure 99. Bamboo flower and fruit. A, Flower of a grass (above) and of a bamboo with six stamens and three styles; ×4. B, Pear-shaped fruit of the bamboo *Melocanna*, with a grass fruit for comparison; ×¼. C, Fruit of *Melocanna* (in section) germinating; ×¼. D, Flowering twig of a bamboo, with the grass leaves transforming into bracts around the spikelets; ×½. (B, C, after Roxburgh, 1819.)

277

plants, dependent on free water for the movement of the sperms, had no opportunity, except for the heterosporous *Selaginella*, a few species of which have become adapted to arid conditions. But in deserts, as mentioned in Chapter 9, we find relics that are the more primitive forms of their families of flowering plants. Here perhaps that outlandish, perennating freak called *Welwitschia*, which is a seedling that develops a woody hypocotyl and merely two basally growing leaves like green leathery laminarian fronds, introduces us to a strangely herbaceous relic of the precursors of the flowering plants. Here survive the branching palm *Hyphaene* and the aloe trees of the lily alliance. If only the explorers could be aware of the thoughts that such plants engender, our appreciation would grow.

Lastly, for this discursive and necessarily incompetent chapter on the travels of plants, there is the domain of fresh water. Rivers, streams, lakes, and ponds would seem to have been the proper passage for sea plants to have become adapted to rain and so to land. Instead, the study of freshwater plant life reveals specialized land plants that have taken to the fresh water and nothing that is truly primitive for land vegetation [107]. Aquatic liverworts are sterile. Aquatic mosses spore above the surface of the water with the air-dry mechanism of land mosses. Aquatic ferns have not only the sporangial mechanism of land ferns, but their stomata, air spaces, roots, and vascular bundles, in deterioration. Aquatic flowering plants add to these land features fruits, seeds, and flowers pollinated by insect and wind, and the histories of these are aerial. Floating duckweeds are the most reduced and specialized members of the aroids, themselves highly evolved monocotyledons of the forest [108]. No other flowering plants are, in fact, so much simplified, and the key to this may be found in the vast watery recesses of the Amazon valley, where the water lettuce *Pistia*, half aroid and half duckweed, thrives. The stoneworts (*Chara*) are filamentous seaweeds without a sporophytic generation, which never made the land. Even the microscopic single-celled and filamentous green algae have in their thick-walled resting zygote the stamp of desiccation that sealed their progress. Nothing pertinent to the upward struggle of timbered plants emerges from the fresh water.

Fresh water, inadequately aerated and poorly charged with mineral matter compared with the productivity of the soil and the sea, is for plants an inferior station and another refuge where the herbaceous, or reduced and short-lived, forms of alga, fungus, moss, fern, or

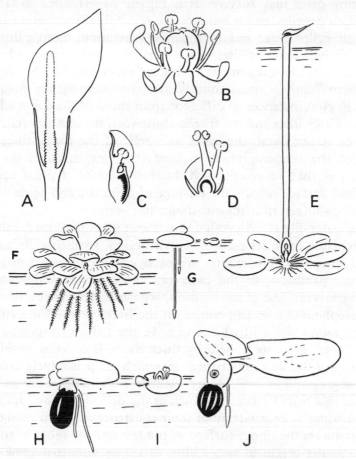

Figure 100. Effect of the freshwater habitat on aroids. A, Normal aroid
inflorescence, in section, as a spike of many minute flowers (female below, male
above), surrounded by a coloured bract and simulating a large flower; × ¼. B,
Bisexual flower of the sweet flag *Acorus*; × 5. C, Small inflorescence of the water
lettuce *Pistia* with one male flower and one female flower; natural size. D,
Minute inflorescence of the duckweed *Lemna*, consisting of one female flower
(represented by an ovary with one ovule) and two male flowers (each
represented by one stamen), surrounded by a minute bract; × 10. E, Malayan
Cryptocoryne griffithii with its long tubular bract opening above the surface of
the water, the base of the tube cut open to show the small spike of flowers in
the beetle room; × ⅙. F, Floating plant of *Pistia*; × ⅙. G, Duckweeds *Lemna*
and *Wolffia* (without root); × 3. H, Seedling of *Pistia* and J, that of *Lemna
trisulca*, compared with the minutest of flowering plants *Wolffia* (I); the seeds
open by a small stopper at the narrow end; × 10. (D, I. J, after Hegelmaier,
1868; C, after Klotsch.)

279

flowering plant may survive. It is replete nevertheless so far as its small fare permits, and it imitates the sea with a freshwater plankton of single-celled algae and an attached vegetation, though this vegetation thrives not in boisterous waters of a rocky shore where the seaweed laid the foundations of higher plant life, but by means of absorbent roots in quiet muddy stretches verging on stagnation. Here, in circumstances so different from those of the oarweed, grow the big water-lilies and the fragile stonewort, the one as a relic of the early flowering plant, the other as a relic of the filamentous green algae of the transmigration to land; the one massive, the other minute, as they represent such different epochs of plant life. The unruffled surface reflects the passage of plant life and some of these reflections linger after the substance has gone.

The water-lily is a dicotyledon without secondary thickening. Its size depends upon its primary construction, swollen cells, and enlarged air spaces. The stouter are pachycaul and vegetatively primitive, comparable with the cacti but under water and with large floating leaves. The giant *Victoria*, whose leaves can be seen from the aeroplane like green pennies on the backwaters of the Amazon, will be found when lifted out to have the habit of a pawpaw; the short, stout, erect stem, held by thick roots, is covered closely with the bases of the long-stalked leaves, which are primitively armoured with savage spines [76]. More elegant water-lilies become leptocaulous and truly herbaceous, such as the delicate *Cabomba*, fashionable in aquaria. Every detail of their construction, from vestigial aril to stomata on the upper surface of the leaf and air tubes in the root, points to the origin of water-lilies as land plants that grew in air, though their roots may have been in the marshy edge of the forest. This habit is retained in part by the Chinese lotus (*Nelumbium*). But water-lilies happen to show well what has been the origin of other fresh water aquatics.

The flowers of water-lilies have a construction that places them low, also, in the scale of floral evolution. They are comparable with *Magnolia* flowers, but the floral axis is shortened and the ovary may become syncarpous and inferior. *Nelumbium* shows uniquely how this tendency to sink the ovary in the stem tissue has happened even with the primitive apocarpous flower. In this reproductive respect, the water-lilies reveal themselves as relics of the early pachycaulous variation of floral structure, and one begins to realize that forests were once planted with a much greater variety of terrestrial water-lilies. What, then, is that genus *Barclaya* of undergrowth water-lilies doing in the rain forests of South-East Asia? Is it a pachycaulous herb ancient or modern? You will find that botany today with its

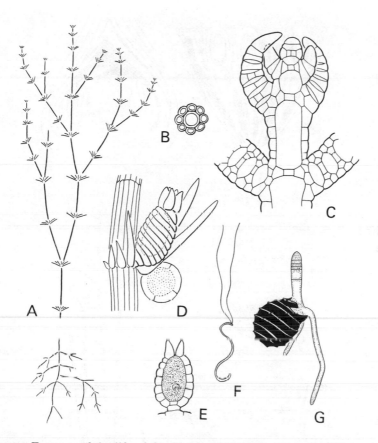

Figure 101. Features of the life of the fresh and brackish water alga *Chara*. A, Plant with verticils of short branches and filamentous root; ×½. B, Stem in section to show the main filament and the filaments of the cortex round it; ×10. C, Apex of the main stem to show the apical cell of the filament, the development of the short branches, the spacing of these by the enlarging central cell of the internode, and the development of the central row of spacing cells in the branches; ×100. D, Part of a short fertile branch with the egg surrounded by spirally grown filaments ending in a crown of short apical cells, and with the round antheridium; ×250. E, Egg with its protective sheath in section; ×250. F, Sperm (antherozoid); ×500. G, Zygote covered by the thick, dark wall formed by the spiral filaments of the egg, and sprouting into a young filamentous plant resembling *Ulothrix* (figure 12); ×250. (After Sachs.)

pressed specimens and photographic experimental records has forgotten how to explore. We need a new generation of alert young biologists eager to enquire into the lives of plants. Laboratories are the studies, not the prisons, of research.

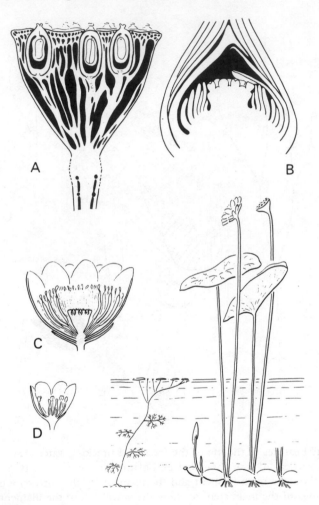

Figure 102. Chinese lotus *Nelumbium*, its creeping stem with swollen internodes, and the slender buttercup-like Brazilian *Cabomba* with feathery underwater leaves; × $\frac{1}{10}$. A, Unique fruit of *Nelumbium* cut open to show the nuts (one-seeded) sunk in the floral axis permeated by air spaces (in black); natural size. B, Young flower-bud of *Nelumbium* cut open to show the young carpels forming on the top of the floral axis as in the *Magnolia* flower (figure 64); × 15. C, Flower of *Nelumbium* with the sterile tips to the stamens; × $\frac{1}{4}$. D, Small flower of *Cabomba* simplified into monocotyledonous form; natural size.

Magnolia has become a forest tree; it does not enter into the reduced and derived vegetation; it is not an aquatic, a desert plant, or a
herb. Botanical books keep it in check as a primitive. Its family is a
hylean success with vast trees supporting abundant life in the modern canopy. The primitive failure is the water-lily. Two more examples with such rather primitive flowers are the buttercups (*Ranun-
culus*) and water plantain (*Alisma*), which, as forest failures, have

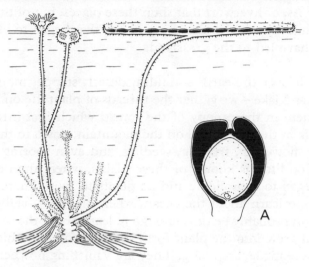

Figure 103. Plant of the very spiny *Victoria amazonica*, its short massive stem
cut longitudinally to show the origin of the copious stout roots from the base of
the leaves; × $\frac{1}{20}$. A, Seed of the water-lily *Nymphaea* containing the small
embryo in the starchy endosperm, and surrounded by a thin aril retaining an
air bubble (in black), which floats and disperses the seed; × 10.

turned aquatic. An example with a simplified flower, spuriously
primitive, is the water milfoil (*Myriophyllum*) related so closely with
the pachycaulous *Gunnera*, like a huge rhubarb. It is the confusion
between the primitive or simplified construction and the aquatic
refuge that has led to the idea that flowering plants and, for that
matter, ferns, mosses, fungi, and green algae of the land have come
by way of fresh water. That alley is blind. But, the vast stretches of
tidal freshwater mud bordering the flowerless forest may have been
where the newcomers gained their massive and spacious setting. It is

283

still the domain of palms, screw pines, and the pachycaulous ancestors of banana, ginger, and aroid. These plants are now ex-
perienced in battle with animals, but their ancestors attracting the
early vegetarians were outdone by them; yet their former presence
is proved by the giant bones of extinct herbivorous reptiles and
mammals that must have eaten something. On what did they live?
We are at the opening question. Zoology reflects botany, but we are
dealing no longer with planktonic pasture engulfed by pseudopodia
and mammoth mouths. We are not thinking of conifer, cycad, tree
fern, moss, or liverwort that shun these places, but of the primitively
and primarily massive flowering plants that began to offer food and
which have led to the grasslands.

From all over the earth – tundra, desert, steppe, mountain, ocean
island, and lake – we gather the threads of plant life on land, and we
trace them in the canopy of the forest, which first fitted plants for
their life in the desert and on the mountain, down to the beginnings
of trees flowering, fruiting, seeding, and even sporing on the river
flats. Too little is known of these places, where the detritus of the
land began to consolidate and its plants to root. There will be time
enough to learn about the ocean, where the plant-cell evolved, and
there will probably be time enough to learn about the seashore where
this cell grew into the plant form. But the forests, which show how
trees were made, are going. They are vanishing nowhere faster than
from the alluvial plains where the vestiges of the last creative phase
of plant life, that prepared the way for the modern world, may survive. The modern mouth is the people's, and theirs the new retalia-
tion. Before machines the forest is defenceless. Human progress is
clearing it with gathering speed to plant crops of quick returns. The
botanist must hurry if he would take the opportunity that a few
brief centuries of his science have revealed; for soon there will be
rice-field to every river-brink. The unmindful tree begot, indeed,
the seed of its own destruction.

Glossary

achene: a small one-seeded, indehiscent fruit.

alga (-ae): a general name for the single-celled plant plankton, seaweeds, and their freshwater allies, variously grouped as Chlorophyceae, Phaeophyceae, Rhodophyceae, etc.

anabolism: the process of building up protoplasm from simple substances.

angiosperm: the flowering plant with ovules contained inside the ovary.

antheridium: the organ producing male gametes.

antherozoids: male gametes.

apical bud: that which terminates a stem.

apocarpous: with the carpels free from each other.

apothecium: the open fungus fructification, as a cup or disc, lined with asci.

archegonium: the female organ of mosses, ferns, and gymnosperms containing the egg inside a cellular wall.

aril: the fleshy envelope more or less surrounding a seed.

ascogonium: the special hypha from which the fertile hyphae develop to produce asci.

ascomycete: a fungus producing asci.

ascus (-i): the reproductive fungus cell (meiotangium) containing usually eight ascospores.

autotrophic: self-feeding from inorganic material.

auxiliary cells: special fusion cells in the development of the carposporophyte of red seaweeds.

auxospore: a cell that restores the original size to the diminishing products of cell-division, as in diatoms.

axil: the junction of a leaf or branch with a larger axis.

axillary bud: the bud in the axil of a leaf (where it joins the stem).

bark: the tissue external to the cambium in a perennial stem or root.

bark-cambium: the layer of cells that produces new bark.

basidiomycete: a fungus producing basidia.

basidium (-a): the reproductive fungus cell (meiotangium) producing usually four spores on the outside.

benthos: attached aquatic plants or animals.

berry: a fleshy, many-seeded indehiscent fruit.

binary fission: the division of a single-celled organism into two daughter-cells.

bract: a reduced leaf on the inflorescence.

bud: a stem-apex covered by developing leaves.

bud-scale: a modified leaf, without lamina, protecting a bud.

calyx: the part of the flower external to the corolla.

cambium: the internal layer of dividing cells that produces secondary xylem and phloem.

capsule: a dehiscent many-seeded fruit derived from a syncarpous ovary.

carpel: the part of the flower that bears the ovules and corresponds with a single leaf.

carpogonium: the female organ of red seaweeds.

carpospore: the non-motile spore of the carposporophyte.

carposporophyte: the post-sexual fruiting stage of a red seaweed, attached to the parent gametophyte.

cauliflory: the state of flowering (and fruiting) from the branches (ramiflory) or trunk.

cell: the unit of protoplasm consisting typically of nucleus, cytoplasm, and, in the plant-cell, photosynthetic pigments, all enclosed by a cell-wall.

chloroplast: the part of the cell containing the chlorophyll.

chromoplast: a chloroplast in which other photosynthetic pigments mask the green colour.

chromosome: one of the set of deeply staining bodies in the nucleus which determine hereditarily cell structure and function.

compound: referring to a leaf, with two or more separate leaflets instead of a single lamina.

conceptacle: a flask-like structure containing reproductive organs.

cone: a reproductive structure of land plants composed of many sporophylls set along a central axis.

conjugation: the union of gametes.

cormophyte: a land plant with shoot and root systems.

corolla: the part of the flower, usually coloured other than green, between the stamens or ovary and the calyx.

cortex: in seaweeds and fungi the tissue external to the central tissue of longitudinal cells; in land plants the tissue of stem or root outside the vascular tissue.

cotyledon: a seed leaf.

cryptogam: a general name for plants not reproduced by seeds.

cuticle: the waterproofing layer on the epidermis of land plants.

cystocarp: a carposporophyte contained in a wall of sterile filaments.

cytoplasm: the part of the protoplasm that is not the nucleus.

dehiscence: opening to let the contents (seeds, spores) escape.

dichotomy: the division of a growing point into two.

dicotyledon (-ous): a flowering plant with two seed leaves (cotyledons).

dioecious: with male and female organs on separate plants.

diploid: with double the chromosome number of the gamete.

dormant: in the state of suspended growth, as undeveloped buds, ungerminated seeds, or resting bulbs.

drupe: a fleshy indehiscent fruit with the one or two seeds enclosed in a stone.

egg: the non-motile female gamete.

epidermis: the outermost layer of cells covering the plant.

epiphyte: a plant that grows on another but not parasitically.

exarillate: without an aril.

eye-spot: the red cytoplasmic structure sensitive to light.

fertilization: the process of conjugation.

fertilization in situ*:* conjugation when the egg is not set free.

filament: a row of cells.

flagellates: cells with flagella but no cell-wall.

flagellum: the thread that is extruded from the cell and moves it.

fructification: the spore-producing structure of a fungus; the structure developed from the ovary of the flower.

gametangium: the organ producing the gametes.

gamete: the sexual cell.

gametophyte: the plant that produces the gametes.

guard-cells: the two cells that bound a stoma.

gymnosperm: a seed plant in which the ovules are not enclosed in carpels or a syncarpous ovary.

hapaxanthic: reproducing once only at the end of the plant's life.

haploid: with the same chromosome number as the gamete.

heterogamy: the union of unequal gametes, that is with recognizably male and female gametes.

heteromorphic: with different forms; in life-cycles, with the sporophyte different in form from the gametophyte.

heterosporous: with male and female spores.

heterotrophic: dependent on organic food made by photosynthetic plants.

homosporous: producing spores alike in appearance and behaviour.

host: the plant on which the parasite grows.

hylea: the primeval forest, particularly the tropical.

hypha (-ae): the non-photosynthetic filaments of fungi and seaweeds.

hypodermis: a special layer of cells immediately internal to the epidermis.

indehiscent: state of a reproductive organ not opening to set free the spores or seeds.

inferior: applied to the ovary of a flower when situated in the stalk below the calyx.

inflorescence: the flowering part of a plant.

integument: applied to ovules and seeds as the covering layer or layers of the sporangium (nucellus).

intercalary: situated between the apex and the base.

internode: the lengthened part of the stem between successive leaves.

isogamy: the union of equal gametes.

isomorphic: having the same form; in life-cycles, with the sporophyte of the same form as the gametophyte.

katabolism: the process of breaking down protoplasm with the liberation of energy and the formation of simple substances.

lamina: the blade of a leaf or leaflet.

leaflet: a unit of a compound leaf.

lenticel: a breathing pore in bark.

leptocaul: with thin or slender primary stem.

lignification: the process of rendering cell-walls woody.

lignin: the substance rendering cell-walls woody.

lomentum: a pod constricted between the seeds.

medulla: the central tissue of a structure, generally referring to the pith, or the central core of seaweeds.

megasporangium (-a): the sporangium containing the megaspores.

megaspore: the female spore.

megasporophyll: the leaf bearing the megasporangia.

meiosis: the nuclear division that halves the chromosome number.

meiotangium: the sporangium or gametangium in which meiosis occurs.

mesophyll: the middle and photosynthetic tissue of a leaf.

metabolism: the physical and chemical processes in protoplasm.

metaxylem: the primary xylem formed after the protoxylem.

micron: one-thousandth of a millimetre.

micropyle: the opening of the integuments of the ovule.

287

microsporangium: the sporangium containing microspores.

microspores: the male spores.

microsporophyll: the leaf bearing the microsporangia.

mitochondria: cytoplasmic structures concerned with enzyme activity.

mitosis: ordinary nuclear division in which the chromosomes are duplicated.

monocotyledon: with one seed leaf.

monoecious: with male and female organs on the same plant.

mother-cell: the cell that gives rise to a particular structure or particular reproductive units.

multicellular: composed of many cells.

multiple fission: the division of a cell into many reproductive units.

mycelium: the vegetative part of a fungus.

mycorrhiza: the non-parasitic association of fungus hyphae with roots.

nectary: a gland secreting sugary nectar.

negative: in conjugation to designate a sex the nature of which is not evident.

neoteny: the process of fulfilling a function in an imperfect or young state.

node: the part of a stem where a leaf is attached.

nucleolus: a special body in the nucleus.

nucleus: the part of the protoplasm that contains the chromosomes.

nucellus: the central body of the ovule equivalent to the megasporangium.

nullipores: calcareous red seaweeds.

nut: a large, woody indehiscent fruit with one or two seeds.

obovate: with egg-shaped outline broadest towards the apex.

oogamy: conjugation between sperms and eggs.

oogonium: the organ producing the egg or eggs.

oosphere: the egg.

ovary: the central part of the flower containing the ovules.

ovate: with egg-shaped outline broadest towards the base.

overtopping: when one branch of a dichotomy grows more than the other.

pachycaul: with thick or massive primary construction.

palisade: referring to the layer of columnar photosynthetic cells in the leaf.

palmate: lobed or divided into fingers, as when the leaflets of a compound leaf are set at the end of the stalk.

parasite: an organism that obtains its food from another (the host).

parenchyma: tissue composed of cells dividing in all directions; false parenchyma, see pseudoparenchyma.

perithecium: a small flask-shaped structure, containing asci.

petal: a segment of the corolla equivalent to a leaf.

phanerogam: a general name for seed plants.

phellogen: the bark cambium.

phloem: the tissue in land plants that conducts organic food material.

photosynthesis: the making of sugar through the action of sunlight.

phyllotaxy: the manner of leaf arrangement on a stem.

phylum: a major group of plants, such as the green seaweeds, the ferns, or the flowering plants.

phyto-: meaning of plant nature, not animal (zoo-).

pinna (*-ae*)*:* a leaflet; also pinnule.

pinnate: with leaflets set along the stalk of a leaf.

pistil: the ovary of a flower.

pit: a thin area in a thickened cell-wall.

plankton: organisms suspended freely in water.

288

plurilocular: referring to a sporangium or gametangium composed of many cells, each producing one zoospore or gamete.

positive: (in reproduction) of opposite sex to negative.

prothallus: the fern gametophyte.

protoplasm: the living contents of a cell.

protoxylem: the first-formed xylem.

pseudoparenchyma: filamentous tissue compacted to resemble parenchyma.

rays: radial strands of cells in wood and phloem.

saprophyte: a plant that lives on dead remains of plants or animals, as a fungus.

secondary: meaning of later construction than primary; secondary thickening, that produced by the cambium.

seed: the product of the ovule containing an embryo sporophyte.

seed fern: extinct fern-like plants that bore seeds on the fronds.

seed leaf: see carpel.

sepal: a segment of the calyx, equivalent to a leaf.

septum (-a): the wall separating two cells of a filament or hypha; the partition inside a structure; septate, provided with septa.

simple: referring to a leaf with single blade, as opposed to the compound leaf with leaflets.

somatic: belonging to the vegetative part of a plant as opposed to the reproductive.

sorus: a cluster of sporangia.

spermatium (-a): the non-motile male cell of red seaweeds; also used for similar cells in fungi.

sporangium (-a): the organ producing spores.

spore: a non-motile and asexual reproductive cell (see zoospore).

sporophyll: a leaf bearing sporangia.

sporophyte: a plant that produces spores or zoospores.

sterigma (-ata): the spike on which a fungus spore is borne.

stichidium: a small structure (lobe or branch) bearing tetrasporangia.

stigma: the part of the style that receives the pollen.

stoma (-ata): the microscopic air pores in the epidermis of land plants.

style: the process on the ovary that bears the stigma.

superior: applied to the ovary of a flower when placed above the calyx.

super-sporangium: a structure containing tetrasporangia and acting as a sporangium, such as the sporangia of ferns and seed plants.

sutures: lines of dehiscence in fruits.

symbiosis: the living together of different species of organism.

syncarpous: with the ovary as a single box containing the one or more ovules, and not identifiable with a carpel; the ovule-bearing tube intercalated below a whorl of carpel initials.

tetrasporangium: the sporangium (meiotangium) containing four spores.

tetraspore: one of the four spores in a tetrasporangium.

thallophyte: a plant that absorbs its food supply over its growing surface, as a seaweed or fungus mycelium (in contrast with a cormophyte).

thallus: the body of a thallophyte.

tracheid: a lignified water-conducting cell.

transpiration: the passage of water through a land plant.

trichogyne: the hair on the female organ of the red seaweeds that receives the male gametes; used also for similar structures in ascomycetes.

triploid: a cell with three times as many chromosomes as the gamete.

unicellular: made of one cell.

unilocular: referring to a sporangium or gametangium made of one mother-cell, which forms many spores or gametes by multiple fission.

vacuole: a watery space in protoplasm, generally in the cytoplasm.

vascular bundle: the conducting strand of a stem or leaf in ferns and seed plants.

vessel: a row of lignified water-conducting cells rendered continuous by the absorption of the cross-walls.

viviparous: bringing forth a young many-celled embryo ready to grow on its own.

xerophyte: a plant accustomed to live under conditions of drought.

xylem: the lignified water-conducting tissue.

zoospore: a motile spore with one or more flagella.

zygomorphic: with one plane of symmetry, generally referring to flowers with upper or lower lips or both.

zygote: the diploid cell formed by the union of two gametes.

Chapter References

Chapter 1

1. Concerning plankton in general, see CHURCH (1919a,e), HARVEY, H. W. (1928, 1942, 1945), HARDY (1956)
   Concerning the ocean, see CARRINGTON (1960)
2. FOGG (1953), HILL & WHITTINGHAM (1955)
3. PRESTON (1952)
4. BAINBRIDGE, R. (1957) The size, shape, and density of marine phytoplanktonic concentrations. *Biological Reviews* **32**: 91–115 (The University Press: Cambridge)

Chapter 2

5. FRITSCH (1935, 1945), GRASSÉ (1952)
6. FAURÉ-FREMIET, E. (1961) Cils Vibratiles et Flagelles. *Biological Reviews* **36**: 464–536 (The University Press: Cambridge)
7. Diatoms: HENDY (1937), HUSTEDT (1930, 1959), KARSTEN (1928), SCHONFELD (1930). Peridineans: LEBOUR (1925), LINDEMANN (1928). Coccolithophores: SCHILLER (1930)
8. COLLIER, A. & MURPHY, A. (1962) Very small diatoms. *Science* **36**: 780–1 (The American Association for the Advancement of Science: Washington, D.C.)

Chapter 3

9. CORNISH (1934), KING (1959)
10. YENDO, K. (1919) A Monograph of the genus Alaria. *Journal of the College of Science, Imperial University of Tokyo* **43**: 1–145. The University, Tokyo. MIYABE, K. (1902) On the Laminariaceae of Hokkaido. English translation. *Journal of the Sapporo Agricultural College* **1** (1957)
11. Concerning seaweeds in general, see CHAPMAN (1962), FOTT (1959), FRITSCH (1935, 1945), OLTMANNS (1922), SMITH, G. (1951), TIFFANY (1958), TILDEN (1935). Concerning seaweed floras, see HARVEY, W. H. (1846–1851, 1851–8, 1858–63), NEWTON (1931), TAYLOR (1957, 1960), GAYRAL (1958)
12. CHURCH (1919a,b), PAPENFUSS (1955)

Chapter 4

13. THOMPSON, D'ARCY (1942)
14. Concerning seaweed making, see the learned dissertation of CHURCH (1920a)
15. GARDINER (1931), SETCHELL, W. A. (1929) Nullipore reef control.

Proceedings of the Fourth Pacific Science Congress, 265. LADD, H. S. (1961) Reef building. *Science* **134**: 703–15 (The American Association for the Advancement of Science: Washington, D.C.)

Chapter 5

16. Concerning the preservation of nature, see HEIM (1956)
17. Shore habits: YONGE (1949). CHAPMAN, V. J. (1957) Marine algal ecology. *The Botanical Review* **23**: 320–50 (The New York Botanical Garden). SOUTHWARD, A. J. (1958) Zonation of plants and animals on rocky sea-shores. *Biological Reviews* **33**: 137–77. (The University Press: Cambridge)
18. Concerning coastal geography, see JOHNSON (1919), STEERS (1953), GUILCHER (1958), WILLIAMS (1960)
19. YONGE, C. M. (1940) The biology of reef-building corals. *The Great Barrier Reef Expedition 1928–29* **1**: no. 13 (The British Museum (Natural History): London)
20. For green seaweeds, see PRINTZ (1927)
21. Concerning phyllotaxy, see CROIZAT, L. (1960) *Principia Botanica* **1**: chapt. 7, p. 633 (Wheldon and Wesley Ltd: Codicote, Hitchin, England)
22. WINGE, O. (1923) The Sargasso Sea, its boundaries and vegetation. *Reports of the Danish Oceanographic Expedition 1908–1910* **3**: no. 2 (Copenhagen). PARR, A. E. (1939) Quantitative observations on the pelagic sargassum vegetation of the western North Atlantic. *Bulletin of the Bingham Oceanographic Collection Peabody Museum of Natural History* **6**: 1–94 (Yale University: New Haven, Conn.)
23. WOMERSLEY, H. B. S. & NORRIS, R. E. (1959) A free-floating marine red alga. *Nature* no. 811, p. 828 (Macmillan and Co. Ltd.: London)
24. SETCHELL, W. A. (1932) Macrocystis and its holdfasts. *The University of California Publications in Botany* **16**: 445–92 (San Francisco)
25. CHURCH (1919c)

Chapter 6

26. CHURCH (1919b), 38, 41
27. For many details on the structure and reproduction of red seaweeds, see KYLIN (1956), FELDMANN-MAZOYER (1940)
28. PRINGSHEIM, E. G. (1949) The relationship between bacteria and myxophyceae (cyanophyceae). *Bacteriological Reviews* **13**: 51–98 (Baltimore)

Chapter 7

29. KENDREW (1961)
30. LUBBOCK (1892)
31. For the world treatment of mosses, see RUHLAND (1924), VERDOORN (1932)
32. Fossil plants: ANDREWS (1961), ARNOLD (1947), BOWER (1935), SCOTT (1924), SEWARD (1931), WALTON (1953). For the synchronization of record of fossil plants and animals, see OAKLEY & MUIR-WOOD (1948)
33. DALLIMORE & JACKSON (1931)
34. WOMERSLEY, J. S. (1958) The araucaria forests of New Guinea. *Proceedings of the Symposium on Humid Tropics Vegetation, Tjiawi (Indonesia)* (UNESCO Science Co-operation Office for South-East Asia: Djakarta)
35. CHAMBERLAIN (1919). ARNOLD, C. A. (1953) Origin and relationship of the cycads. *Phytomorphology* **3**: 51–66 (The International Society of Plant Morphologists: The University, Delhi)
36. FAVRE-DUCHARTRE, M. (1958) Ginkgo, an oviparous plant. *Phytomorphology* **8**: 377–90 (The International Society of Plant Morphologists: The University, Delhi)
37. For plant geography, see CROIZAT (1958, the most profound work), ENGLER & DRUDE (1896–1928), GOOD (1953), KARSTEN & SCHENCK (1904–44), SCHIMPER (1903, 1935), WALTER (1962)

Chapter 8

38. For a recent symposium on leaves, see MILTHORPE (1956)
39. For comprehensive works on the anatomy of land plants, see EAMES (1961), ESAU (1953, 1960), HABERLANDT (1914)
40. McCURRACH (1960)
41. YVON, F. DUROCHER (1947) Seychelles Botanical Treasure. *La Revue Agricole de l'Ile Maurice* **26**: 69–87 (Mauritius)
42. KUMAZAWA, M. (1961) Studies on the vascular course in maize plant. *Phytomorphology* **11**: 128–139 (The International Society of Plant Morphologists: The University, Delhi)
43. BROWN, S. A. (1961) Chemistry of lignification. *Science* **134**: 305–313 (The American Association for the Advancement of Science: Washington, D.C.)
44. CRAFTS, A. S., CURRIER, H. B. & STOCKING, C. R. (1949) *Water in the physiology of plants* 168–72 (Chronica Botanica Co.: Waltham, Mass.). SCHOLANDER, P. F., HEMMINGSEN, E. & GAREY, W. (1961) Cohesive lift of sap in the rattan vine. *Science* **134**: 1835–8 (The American Association for the Advancement of Science: Washington, D.C.)
45. CHEADLE, V. I. (1953) Independent origin of vessels in the monocotyledons and dicotyledons. *Phytomorphology* **3**: 23–44 (The International Society of Plant Morphologists: The University, Delhi)
46. JANE, F. W. (1956) The Structure of wood (A. and C. Black: London). FAHN, A. (1962) *Plant anatomy* (The Hebrew Univer-

sity: Jerusalem; in Hebrew, but with excellent illustrations)

47. SPRUCE (1908) **1**: 20–3
48. WHITMORE, T. C. (1962) Studies in systematic bark-morphology, I, II. *The New Phytologist* **61**: 191–220 (Blackwell Scientific Publications: Oxford)
49. EDGAR (1958) 317–55

50. For an introduction to plant physiology, see MEYER & ANDERSON (1960), MILLER (1953). For an advanced treatise, see BONNER & GALSTON (1952), KRAMER & KOZLOWSKI (1960), MEYER & ANDERSON (1952), THOMAS, RANSON & RICHARDSON (1956), WITHROW (1959)

Chapter 9

51. For an introduction to tropical trees, see CORNER (1952)
52. For botanical books on tropical forest, see ALLEN (1956), AUBERT DE LA RUE, BOURLIÈRE & HARROY (1954), BEWS (1925), DECKER (1936), PURI (1960), RICHARDS (1952)
53. WILLIS (1948), one of the most useful books in botany
54. HILLEBRAND, W. F. (1888) *Flora of the Hawaiian Islands* 235
55. ARBER (1961)
56. HOLTTUM, R. E. (1955) Growthhabits of monocotyledons. *Phyto-morphology* **5**: 399–413 (The International Society of Plant Morphologists: The University, Delhi)
57. AUDUS (1959)
58. EAMES, A. J. (1953) Neglected morphology of the palm leaf. *Phytomorphology* **3**: 172–90. PERIASAMY, K. (1962) Morphological and ontogenetic studies in palms. *Phytomorphology* **12**: 54–64 (The International Society of Plant Morphologists: The University, Delhi)

Chapter 10

59. CHURCH (1924–5)
60. CROCKER & BARTON (1953)
61. RIDLEY (1930)
62. CHURCH (1919b) 4
63. INGOLD (1939)
64. For ferns, see Bower (1923–8), VERDOORN (1938), HOLTTUM, R. E. & SEN, U. (1961) Morphology and classification of the tree ferns. *Phytomorphology* **11**: 406–420 (The International Society of Plant Morphologists: The University, Delhi)
65. MAHESHWARI (1950)
66. RANGA SWAMY, N. S. (1961) Experimental studies on female reproductive structures of *Citrus microcarpa* Bunge. *Phytomorpho-logy* **11**: 109–27 (The International Society of Plant Morphologists: The University, Delhi)
67. CHAMBERLAIN (1957)
68. EAMES (1936), SMITH, G. (1938)
69. PURI, V. & GARG, M. L. (1953) A contribution to the anatomy of the sporocarp of *Marsilea minuta* L. *Phytomorphology* **3**: 190–209 (The International Society of Plant Morphologists: The University, Delhi). MEEUSE, A. D. J. (1961) Marsileales and Salviniales – 'Living Fossils'. *Acta Botanica Neerlandica* **10**: 257–60 (North Holland Publishing Co.: Amsterdam)

Chapter 11

70. KORIBA, K. (1958) On the periodicity of tree growth in the tropics. *Gardens' Bulletin, Singapore* **17**: 11–82 (Botanic Gardens, Singapore)

71. For botanical accounts of flowers, see CHURCH (1908), MCLEAN & IVIMEY-COOK (1956, vol. 2). For an introduction to tropical flowers, see MENNINGER (1962)

72. FAHN, A. (1953) The topography of the nectary in the flower and its phylogenetic trend. *Phytomorphology* **3**: 424–6 (The International Society of Plant Morphologists: The University, Delhi)

73. ROBERTSON, C. (1904) The structure of the flowers and the mode of pollination of the primitive angiosperms. *The Botanical Gazette* **37**: 294–8 (The University of Chicago, Illinois)

74. MEEUSE (1961), the best introduction to animal and flower relations. V. DER PIJL, L. (1953). On the flower biology of some plants from Java, with general remarks on fly-traps. *Annales Bogorienses* **1**: 77–99 (Bogor, Indonesia). V. DER PIJL, L. (1954) Xylocarpa and flowers in the tropics. *Proceedings of the Royal Academy of Sciences, Amsterdam,* Ser. C **57**: 413–23, 541–62. V. DER PIJL, L. (1960–1961) Ecological aspects of flower evolution. I, Phyletic evolution. *Evolution* **14**: 403–16. II, Zoophilous flower classes. *Evolution* **15**: 44–59 (The Society for the Study of Evolution: Lancaster, Pennsylvania)

75. KNUTH (1906–9)

76. DECKER (1936) pp. 46–53

77. PICKENS, A. L. (1936) Steps in the development of the bird-flower. *Condor* **38**: 150–4 (Cooper Ornithological Club: Santa Clara, California)

78. GRANT, V. (1950) The protection of the ovules in flowering plants. *Evolution* **4**: 179–201 (The Society for the Study of Evolution: Lancaster, Pennsylvania)

79. WERTH, E. (1915) Kurzer ueberblick ueber die gesamtfrage der ornithophilie. *Engler Botanische Jahrbuch* **53**: 314–78 (Wilhelm Engelmann: Leipzig)

80. V. DER PIJL, L. (1956) Remarks on pollination by bats. *Acta Botanica Neerlandica* **5**: 135–44 (North Holland Publishing Co.: Amsterdam)

Chapter 12

81. CORNER, E. J. H. (1951) The leguminous seed. *Phytomorphology* **1**: 117–50 (The International Society of Plant Morphologists: The University, Delhi)

82. CORNER (1949, 1953–4, 1954)

83. WALLACE (1913) p. 57

84. V. DER PIJL, L. (1957) The dispersal of plants by bats (chiropterochory). *Acta Botanica Neerlandica* **6**: 291–315 (North Holland Publishing Co.: Amsterdam)

85. TAKHTAJAN (1959), ZIMMERMANN (1959). ARBER, E. A. N. & PARKIN, J. (1907) The origin of angiosperms. *The Journal of the Linnean Society, Botany* **38**: 29–80 (The Linnean Society of London)

Chapter 13

86. For general accounts of fungi, see INGOLD (1953), RAMSBOTTOM (1953), WAKEFIELD & DENNIS (1950). For the classification of fungi, see AINSWORTH & BISBY (1961), BESSEY (1950), DENNIS (1960), WOLF & WOLF (1947). For the physiology of fungi, see HAWKER (1950), LILLY & BARNETT (1951)

87. CORNER (1950)

88. BULLER (1909–50)

Chapter 14

89. CORNER, E. J. H. (1960) Larger fungi in the tropics. *Transactions of the British Mycological Society* **24**: 357 (The University Press: Cambridge)

90. HEIM, R. (1958) *Termitomyces. Flore iconographique des champignons du Congo*. Fascicule 7 (Jardin Botanique de l'État: Brussels). SINGER, R. (1962) *The agaricales in modern taxonomy*, pp. 432, 466 (Wheldon and Wesley: Codicote, Herts.)

91. CORNER, E. J. H. (1929–30) Studies in the morphology of discomycetes. *Transactions of the British Mycological Society* **14**: 263–91; **15**: 107–34, 332–50 (The University Press: Cambridge)

92. BURGES (1958), JACKS (1959), PARKINSON & WAID (1960)

93. HARLEY (1959)

94. GREGORY (1961)

95. GARRETT (1956)

96. DUNCAN (1959), HALE (1961), SMITH, A. L. (1921), ZAHLBRUCKNER (1926), ZAHLBRUCKNER & KEISSLER (1930–60)

97. CHURCH (1920b), (1921a,b)

Chapter 15

98. HUMBOLDT, A. VON (1849) *Ansichten der Natur* **1**: 13 (Ed. 3, Stuttgart; Ed. 1, 1808)

99. RAUNKIAER (1934)

100. HUMBOLDT (1807)

101. V. STEENIS, C. G. G. J. (1958) Ecology of mangrove. *Flora Malesiana* Ser. 1 **54**: 431–45 (Noordhoff-Kolff N.V.: Djakarta)

102. CHAPMAN (1960)

103. V. STEENIS, C. G. G. J. (1954) *Homo Destruens* (Noordhoff-Kolff N.V.: Amsterdam; in Dutch)

104. ARBER (1934), BEWS (1929)

105. SEIFRIZ, W. (1950) Gregarious flowering of Chusquea. *Nature* **165**: 635 (Macmillan and Company Ltd.: London)

106. BUXBAUM, F. (1951) *Grundlagen und methoden einer erneuerung der systematik der höheren pflanzen* (Springer Verlag: Vienna)

107. ARBER (1920)

108. MAHESHWARI, S. C. (1954) The embryology of Wolffia. *Phytomorphology* **4**: 355–65 (The International Society of Plant Morphologists: The University, Delhi)

Selected Bibliography

AINSWORTH, G. C. & BISBY, G. R. (1961) *Dictionary of the fungi.* 5th Ed. (Commonwealth Mycological Institute: Kew)

ALLEN, P. H. (1956) *The rain forests of Golfo Dulce* (University of Florida Press: Gainesville)

ANDREWS, H. N. (1961) *Studies in palaeobotany* (John Wiley and Sons Ltd.: New York and London)

ARBER, A. (1920) *Water plants: a study of aquatic angiosperms* (University Press: Cambridge)

ARBER, A. (1934) *The Gramineae: a study of cereal, bamboo, and grass* (University Press: Cambridge)

ARBER, A. (1961) *Monocotyledons* (Reprinted by J. Cramer-Weinheim, Hafner Publishing Co.: New York; Wheldon and Wesley: Codicote, Herts.)

ARNOLD, C. A. (1947) *An introduction to palaeobotany* (McGraw-Hill Book Co. Inc.: New York)

AUBERT DE LA RUE, BOURLIERE, F. & HARROY, J. P. (1954) Tropiques (la nature tropicale). Horizons de France, Paris

AUDUS, L. J. (1959) *Plant growth substances.* 2nd ed. Plant Science Monographs, Leonard Hill Ltd.: London)

BESSEY, E. A. (1950) *Morphology and taxonomy of fungi* (The Blakiston Co.: Toronto)

BEWS, J. W. (1925) *Plant forms and their evolution in South Africa* (Longmans, Green and Co. Ltd.: London)

BEWS, J. W. (1929) *The world's grasses* (Longmans, Green and Co. Ltd.: London)

BONNER, J. & GALSTON, A. W. (1952) *Principles of plant physiology* (W. H. Freeman: San Francisco)

BOWER, F. O. (1923–28) *The ferns (Filicales).* 3 vols. (University Press: Cambridge)

BOWER, F. O. (1935) *Primitive land plants* (Reprinted (1959) by the Hafner Publishing Co.: New York)

BULLER, A. H. R. (1909–50) *Researches on fungi.* 7 vols. (Vol. 1–6, Longmans, Green and Co. Ltd.: London. Vol. 7, The Royal Society of Canada: University of Toronto Press)

BURGES, A. (1958) *Micro-organisms in the soil* (Hutchinson: University Library, London)

CARRINGTON, R. (1960) *A biography of the sea* (Chatto and Windus Ltd.: London)

CHAMBERLAIN, C. J. (1919) *The living cycads* (University of Chicago Press)

CHAMBERLAIN, C. J. (1957) *Gymnosperms, structure and evolution.* (Reprinted by Johnson Reprint Corporation: New York. 1934; University of Chicago Press)

CHAPMAN, V. J. (1962) *The algae* (Macmillan and Co. Ltd.: London)

CHAPMAN, V. J. (1960) *Salt marshes and salt deserts of the world* (Leonard Hill Ltd.: London)

CHURCH, A. H. (1908) *Types of floral mechanism* (Clarendon Press: Oxford)

CHURCH, A. H. (1919a) *The building of an autotrophic flagellate.* Oxford Botanical Memoirs no. 1 (Clarendon Press: Oxford)

CHURCH, A. H. (1919b) *Thalassiophyta and the subaerial transmigration.* Oxford Botanical Memoirs no. 3 (Clarendon Press: Oxford)

CHURCH, A. H. (1919c) Weighing moorings. *Journal of Botany* **57**: 35–7 (Taylor and Francis Ltd.: London)

CHURCH, A. H. (1919d) The phaeophycean zoid. *Journal of Botany* **57**: supplement II, 1–7 (Taylor and Francis Ltd.: London)

CHURCH, A. H. (1919e) Plankton-phase and plankton-rate. *Journal of Botany* **57**: supplement III, 1–8 (Taylor and Francis Ltd.: London)

CHURCH, A. H. (1919f) Historical review of phaeophyceae. *Journal of Botany* **57**: 265–73 (Taylor and Francis Ltd.: London)

CHURCH, A. H. (1919g) Historical review of the florideae (red seaweeds). *Journal of Botany* **57**: 297–304, 329–34 (Taylor and Francis Ltd.: London)

CHURCH, A. H. (1920a) *The somatic organisation of the phaeophyceae.* Oxford Botanical Memoirs no. 10 (Clarendon Press: Oxford)

CHURCH, A. H. (1920b) The lichen symbiosis. *Journal of Botany* **58**: 213–19, 262–7 (Taylor and Francis Ltd.: London)

CHURCH, A. H. (1921a) The lichen as transmigrant. *Journal of Botany* **59**: 7–13, 40–6 (Taylor and Francis Ltd.: London)

CHURCH, A. H. (1921b) The lichen life-cycle. *Journal of Botany* **59**: 139–47, 164–170, 197–202, 216–21 (Taylor and Francis Ltd.: London)

CHURCH, A. H. (1924–25) Reproductive mechanism in land-flora. *Journal of Botany* **62**: 108–12, 139–42, 209–14, 268–75; **63**: 15–20, 78–85, 132–8, 193–8; **64**: 33–40, 99–103, 132–6, 172–8, 211–15, 234–40, 257–62, 307–10, 332–6 (Taylor and Francis Ltd.: London)

CORNER, E. J. H. (1949) The durian theory or the origin of the modern tree. *Annals of Botany* New Series **13**: 367–414 (Clarendon Press: Oxford)

CORNER, E. J. H. (1953–54) The durian theory extended, Parts I, II, III. *Phytomorphology* **3**: 465–76; **4**: 152–65, 263–74 (The International Society of Plant Morphologists: The University, Delhi)

CORNER, E. J. H. (1950) *A monograph of clavaria and allied genera.* Annals of Botany Memoirs no. 1 (University Press: Oxford)

CORNER, E. J. H. (1952) *Wayside trees of Malaya,* 2nd ed. 2 vols. (Government Printer: Singapore)

CORNER, E. J. H. (1954) *The evolution of tropical forest. Evolution as a process.* Ed. by Huxley, J. S., Hardy, A. C. & Ford, E. B. 34–46 (Allen and Unwin: London)

CORNISH, V. (1934) *Ocean Waves* (University Press: Cambridge)

CROCKER, W. & BARTON, L. V. (1953) *Physiology of seeds* (Chronica Botanica Co.: Waltham, Mass.)

CROIZAT, L. (1958) *Panbiogeography.* 3 vols. (published by the author: Caracas)

DALLIMORE, W. & JACKSON, A. BRUCE (1931) *A handbook of coniferae (including Gingkoaceae)*. 2nd ed. (Edward Arnold Ltd.: London)

DECKER, J. S. (1936) *Aspectos biologicos da flora brasileira* (Rotermund: São Leopoldo, Rio Grande do Sul, Brazil)

DENNIS, R. W. G. (1960) *British cup-fungi and their allies*. (The Ray Society. Bernard Quaritch Ltd.: London)

DUNCAN, U. K. (1959) *A guide to the study of lichens* (T. Buncle: Arbroath)

EAMES, A. J. (1936) *Morphology of vascular plant. Lower plants* (McGraw-Hill Book Co. Inc.: New York)

EAMES, A. J. (1961) *Morphology of the angiosperms* (McGraw-Hill Book Co. Inc.: New York)

EDGAR, A. T. (1958) *Manual of rubber-planting* (The Incorporated Society of Planters: Kuala Lumpur)

ENGLER, A. & DRUDE, O. (1896–1928) *Die vegetation der erde*. 14 vols. (Wilhelm Engelmann: Leipzig)

ESAU, K. (1953) *Plant anatomy* (John Wiley and Sons Ltd.: New York; Chapman and Hall Ltd.: London)

ESAU, K. (1960) *Anatomy of seed plants* (John Wiley and Sons Ltd.: New York; Chapman and Hall Ltd.: London)

FELDMANN-MAZOYER, G. (1940) *Recherches sur les céramiacées de la mediterranée occidentale* (Imprimerie Minerva: Algiers)

FOGG, G. E. (1953) *The metabolism of the algae* (Methuen and Co. Ltd.: London)

FOTT, F. B. (1959) *Algenkunde* (Gustav Fischer: Jena)

FRITSCH, F. E. (1935, 1945) *The structure and reproduction of the algae*. 2 vols. (University Press: Cambridge)

GARDINER, S. J. (1931) *Coral reefs and atolls* (Macmillan and Co. Ltd.: London)

GARRETT, S. D. (1956) *Biology of root-infecting fungi* (University Press: Cambridge)

GAIRAL, P. (1958) *Algues de la côte Atlantique Marocaine* (Rabat, Morocco)

GOOD, R. (1953) *The geography of the flowering plants*. 2nd ed. (Longmans, Green and Co. Ltd.: London)

GRASSÉ, P. P. (1952) *Traité de Zoologie*. Tome 1. *Protozoaires. Flagelles* (Masson et Cie: Paris)

GREGORY, P. H. (1961) *The microbiology of the atmosphere*. Plant Science Monographs (Leonard Hill Ltd.: London)

GUILCHER, A. (1958) *Coastal and submarine morphology*. Translated from the French by SPARKS, B. W. & KNEESE, R. H. W. (Methuen and Co. Ltd.: London)

HABERLANDT, G. (1914) *Physiological plant anatomy*. Translated from the 4th German ed. by DRUMMOND, M. (Macmillan and Co. Ltd.: London)

HALE, M. E. (1961) *Lichen handbook. A guide to the lichens of eastern North America* (Smithsonian Institution: Washington, D.C.)

HARDY, A. C. (1956) The open sea. *The New Naturalist* (Wm. Collins Sons and Co. Ltd.: London)

HARLEY, J. (1959) *The biology of mycorrhiza*. Plant Science Monographs (Leonard Hill Ltd.: London)

HARVEY, H. W. (1928) *Biological chemistry and physics of sea water* (University Press: Cambridge)

299

HARVEY, H. W. (1942) Production of life in the sea. *Biological Reviews* **17**: 221–46 (University Press: Cambridge)

HARVEY, H. W. (1945) *Recent advances in the chemistry and biology of sea water* (University Press: Cambridge)

HARVEY, W. H. (1846–51) *Phycologia brittanica or history of British seaweeds*. 4 vols. (Reeve Brothers: London)

HARVEY, W. H. (1851–58) *Nereis Boreali-Americana* (The Smithsonian Institution: Washington, D.C.)

HARVEY, W. H. (1858–63) *Phycologia australica, or a history of Australian seaweeds*. 5 vols. (Lovell Reeve: London)

HAWKER, L. E. (1950) *Physiology of fungi* (University of London Press: London)

HEIM, R. (1956) *Derniers refuges*. The International Union for the Conservation of Nature, Morge, Switzerland (Elsevier: Paris)

HENDY, N. I. (1937) The plankton diatoms of the Southern Seas. *'Discovery' Reports* **16**: 151–364. The National Institute of Oceanography (University Press: Cambridge)

HILL, R. & WHITTINGHAM, C. P. (1955) *Photosynthesis*. Methuen's Monographs on Biochemical Subjects (Methuen and Co. Ltd.: London)

HUMBOLDT, A. v. (1807) *Voyage de Humboldt et Bonpland*, vol. 1. *Essai sur la géographie des plantes* (Fr. Schoell: Paris)

HUSTEDT, F. (1930, 1959) *Die Kieselalgen (Diatomaceae)*. Rabenhorst's *Kryptogamen Flora*, Bd. 7 (Akademische Verlagsgesellschaft: Leipzig)

INGOLD, C. T. (1939) *Spore discharge in land plants* (Clarendon Press: Oxford)

INGOLD, C. T. (1953) *Dispersal in fungi* (Clarendon Press: Oxford)

JACKS, G. V. (1959) *Soil* (Thomas Nelson and Sons Ltd.: London)

JOHNSON, D. W. (1919) *Shore-processes and shoreline development* (Chapman and Hall Ltd.: London)

KARSTEN, G. (1928) Diatomeae. Engler, A., *Die Natürlichen Pflanzenfamilien*. 2nd ed. Bd. 2 (Wilhelm Engelmann: Leipzig)

KARSTEN, G. & SCHENCK, H. (1904–44) *Vegetationsbilder*. 26 vols. (Gustav Fischer: Jena)

KENDREW, W. G. (1961) *The climates of the continents*. 5th ed. (Clarendon Press: Oxford)

KING, C. A. M. (1959) *Beaches and coasts* (Edward Arnold Ltd.: London)

KNUTH, P. (1906–9) *Handbook of flower pollination*. 3 vols. Translated from the German by DAVIS, R. A. (University Press: Oxford)

KRAMER, P. J. & KOZLOWSKI, T. T. (1960) *Physiology of trees* (McGraw-Hill Book Co. Inc.: New York)

KYLIN, H. (1956) *Die Gattungen der Rhodophyceen*. (CWK Gleerups Förlag: Lund)

LEBOUR, M. V. (1925) *The dinoflagellates of northern seas* (Marine Biological Association: Plymouth, U.K.)

LILLY, V. G. & BARNETT, H. L. (1951) *Physiology of the fungi* (McGraw-Hill Book Co. Inc.: New York)

LINDEMANN, E. (1928) Peridineae. Engler, A., *Die Natürlichen Pflanzenfamilien*. 2nd ed. Bd. 2 (Wilhelm Engelmann: Leipzig)

LUBBOCK, Sir JOHN (1892) *A contribution to our knowledge of seedlings*. 2 vols. (Kegan Paul Trench Trubner and Co. Ltd.: London)

MAHESHWARI, P. (1950) *An introduction to the embryology of angiosperms* (McGraw-Hill Book Co. Inc.: New York)

MCCURRACH, J. (1960) *Palms of the world* (Harper and Brothers: New York)

MCLEAN, R. C. & IVIMEY-COOK, W. R. (1956) *Textbook of theoretical botany*. 2 vols. (Longmans, Green and Co. Ltd.: London)

MEEUSE, B. J. D. (1961) *The story of pollination* (The Ronald Press Co.: New York)

MENNINGER, E. A. (1962) *Flowering trees of the world* (Hearthside Press Inc.: New York)

MEYER, B. S. & ANDERSON, D. B. (1952) *Plant physiology*. 2nd ed. (D. Van Nostrand Co. Inc.: New York)

MEYER, B. S. (1960) *Introduction to plant physiology* (D. Van Nostrand Co. Inc.: New York)

MILLER, E. V. (1953) *Within the living plant* (The Blakiston Co. Inc.: New York and Toronto)

MILTHORPE, F. L. (1956) *The growth of leaves* (Butterworths Scientific Publications: London)

NEWTON, L. (1931) *A handbook of the British seaweeds* (British Museum (Natural History): London)

OAKLEY, K. P. & MUIR-WOOD, H. M. (1948) *The succession of life through geological time* (British Museum (Natural History): London)

OLTMANNS, F. (1922) *Morphologie und Biologie der Algen*. 3 vols. (Gustav Fischer: Jena)

PAPENFUSS, G. F. (1955) *Classification of the algae. A century of progress in the natural sciences*, 1853–1953, 115–224. California Academy of Sciences, San Francisco

PARKINSON, D. & WAID, J. S. (1960) *The ecology of soil fungi*. An International Symposium (Liverpool University Press)

PASCHER, A. (1918) Ueber diploide Zwerggenerationen bei Phaeophyceen (Laminaria saccharina). *Berichte der Deutschen Botanischen Gesellschaft*. **36**: 246–52 (Gebrüder Borntraeger: Berlin)

PRESTON, R. D. (1952) *The molecular architecture of plant cell-walls* (Chapman and Hall Ltd.: London)

PRINTZ, H. (1927) Chlorophyceae. Engler, A., *Die Natürlichen Pflanzenfamilien*. 2nd ed. Bd. 3 (Wilhelm Engelmann: Leipzig)

PURI, G. S. (1960) *Indian forest ecology*. 2 vols. (Oxford Book and Stationery Co.: New Delhi and Calcutta)

RAMSBOTTOM, J (1953) Mushrooms and toadstools. *The New Naturalist* (Wm. Collins Sons and Co. Ltd.: London)

RAUNKIAER, C. (1934) *The life forms of plants* (Clarendon Press: Oxford)

RICHARDS, P. W. (1952) *The tropical rain forest. An ecological study* (University Press: Cambridge)

RIDLEY, H. N. (1930) *The dispersal of plants throughout the world* (L. Reeve: Ashford, Kent)

RUHLAND, W. (1924) Musci. Engler, A., *Die Natürlichen Pflanzenfamilien*. Bd. 10 (Wilhelm Engelmann: Leipzig)

SCHILLER, J. (1930) Coccolithineae. Rabenhorst's *Kryptogamen Flora*. Bd. 10, abt. 2 (Akademische Verlagsgesellschaft: Leipzig)

THE LIFE OF PLANTS

SCHILLER, J. (1933, 1937) Dinoflagellatae. Rabenhorst's *Kryptogamen Flora*. Bd. 10, abt. 2 (Akademische Verlagsgesellschaft: Leipzig)

SCHIMPER, A. F. W. (1903) *Plant-geography upon a physiological basis.* Translated from the German by FISCHER, W. R. Ed. by GROOM, P. & BALFOUR, I. B. (University Press: Oxford)

SCHIMPER, A. F. W. (1935) *Pflanzengeographie auf physiologischer Grundlage.* 2 vols. 3rd ed. Revised by VAN FABER, F. C. (Gustav Fischer: Jena)

SCHONFELD, H. V. (1930) Bacillariales (Diatomeae). Pascher's *Süsswasser Flora von Deutschland, Oesterreich, und der Schweiz.* Heft. 10 (Gustav Fischer: Jena)

SCOTT, D. H. (1924) *Extinct plants and problems of evolution* (Macmillan and Co. Ltd.: London)

SETCHELL, W. A. (1933) Hong Kong seaweeds, III. Sargassaceae. *The Hong Kong Naturalist Supplement* no. 2, 33–49 (The Newspaper Enterprise Ltd.: Hong Kong)

SEWARD, A. C. (1931) *Plant life through the ages* (University Press: Cambridge)

SMITH, A. L. (1921) *Lichens* (University Press: Cambridge)

SMITH, G. (1938) *Cryptogamic botany.* 2 vols. (McGraw-Hill Book Co. Inc.: New York)

SMITH, G. (1951) *Manual of phycology* (Chronica Botanica Company: Waltham, Mass.)

SPRUCE, R. (1908) *Notes of a botanist on the Amazon and Andes, 1849–1864.* 2 vols. Ed. by WALLACE, A. R. (Macmillan and Co. Ltd.: London)

STEERS, J. A. (1953) The sea coast. *The New Naturalist* (Wm. Collins Sons and Co. Ltd.: London)

TAKHTAJAN, A. (1959) *Die evolution der angiospermen.* Translated into German from the Russian by HÖPPNER, W. (Gustav Fischer: Jena)

TAYLOR, W. R. (1957) *Marine algae of the northeastern coast of North America* (University of Michigan Press: Ann Arbor, Michigan)

TAYLOR, W. R. (1960) *Marine algae of the eastern tropical and subtropical coast of the Americas* (University of Michigan Press: Ann Arbor, Michigan)

THOMAS, M., RANSON, S. L. & RICHARDSON, J. A. (1956) *Plant physiology.* 4th ed. (J. and A. Churchill Ltd.: London)

THOMPSON, D'ARCY W. (1942) *On growth and form.* New ed. (University Press: Cambridge)

THOMPSON, J. MCLEAN (1933) *The theory of scitaminean flowering.* Publications of the Hartley Botanical Laboratories, no. 11, Part 6 (University Press of Liverpool)

THURET, G. (1878) *Etudes phycologiques* (Paris)

TIFFANY, L. H. (1958) *Algae – The grass of many waters* (Blackwell Scientific Publications: Oxford)

TILDEN, J. (1935) *The algae and their life-relations* (University of Minnesota Press)

VERDOORN, F. (1932) *Manual of bryology* (Martinus Nijhoff: The Hague)

VERDOORN, F. (1938) *Manual of pteridology* (Martinus Nijhoff: The Hague)

WAKEFIELD, E. M. & DENNIS, R. W. G. (1950) *Common British fungi* (P. R. Gawthorn: London)

WALLACE, A. R. (1913) *The Malay archipelago* (Macmillan and Co. Ltd.: London)

WALTER, H. (1962) *Die vegetation der erde.* 2 vols. (Gustav Fischer: Jena)

WALTON, J. (1953) *An introduction to the study of fossil plants* (A. and C. Black: London)

WILLIAMS, W. W. (1960) *Coastal changes* (Routledge and Kegan Paul Ltd.: London)

WILLIS, J. C. (1948) *A dictionary of flowering plants and ferns.* 6th ed. (University Press: Cambridge)

WITHROW, R. B. (1959) Photoperiodism and related phenomena in plants and animals. *Proceedings of the Conference on Photoperiodism* (1957) (The American Association for the Advancement of Science: Washington, D.C.)

WOLF, F. A. & WOLF, F. T. (1947) *The fungi.* 2 vols. (John Wiley and Sons Ltd.: New York; Chapman and Hall Ltd.: London)

YONGE, C. M. (1949) The sea shore. *The New Naturalist* (Wm. Collins Sons and Co. Ltd.: London)

ZAHLBRUCKNER, A. (1926) Lichenes. Engler, A., *Die Natürlichen Pflanzenfamilien.* Bd. 8 (Wilhelm Engelmann: Leipzig)

ZAHLBRUCKNER, A. & KEISSLER, K. v. (1930–60) Die flechten (Lichenes). Rabenhorst's *Kryptogamen Flora von Deutschland, Oesterreich, und der Schweiz.* Bd. IX (Akademische Verlagsgesellschaft: Geest und Portig K.-G., Leipzig)

ZIMMERMANN, W. (1959) *Die Phylogenie der Pflanzen* (Gustav Fischer: Stuttgart)

Index

84-3237/MCL

QK   Corner, E. J. H.
45.2   (Edred John Henry)
.C67
1981   The life of plants
c. 2

DATE			
AG1 1 '88			
JUL 20 '88			

84-3237/MCL